Building the New Jerusalem

Mark Swenarton

Building the New Jerusalem

Architecture, housing and politics 1900–1930

Mark Swenarton

IHS BRE Press publications are available from
www.ihsbrepress.com
or
IHS BRE Press
Willoughby Road
Bracknell RG12 8FB
Tel: 01344 328038
Fax: 01344 328005
Email: brepress@ihs.com

Requests to copy any part of this publication should be
made to the publisher:
IHS BRE Press
Garston, Watford WD25 9XX
Tel: 01923 664761
Email: brepress@ihs.com

Printed on paper sourced from responsibly managed
forests

Cover design by Hayley Bone

EP 82

First published 2008

ISBN 978-1-84806-024-1

To the memory of
Mavis Swenarton

CONTENTS

CD CONTENTS

The accompanying CD contains scanned copies of the following documents.

FIGURES

Introduction

And I saw an angel come down from heaven, having the key of the bottomless pit and a great chain in his hand. And he laid hold on the dragon, that old serpent which is the Devil, and Satan, and bound him a thousand years.... And I saw a new heaven and a new earth: for the first heaven and the first earth were passed away; and there was no more sea. And I John saw the holy city, new Jerusalem, coming down from God out of heaven.

(Revelation, chapters 20–21)

The period 1900-1930 was the one in which the architecture, housing and politics of the modern world were formed. A time traveller from the twenty-first century arriving in 1900 Europe would find themselves in a foreign land: no political consensus over the state's responsibility for welfare; no widespread provision of social housing by the state; no 'modern architecture' in the form that we know today and no concept of housing as a major part of architecture. Wind the clock forward by 30 years and the landscape would be familiar: widespread state involvement in the provision of housing for the working class; social democracy established as one of (if not the) dominant political formations; a theory and practice of modern architecture that, in its essentials, is still with us today, with housing seen as a major component of the discipline of architecture.

Nor, of course, was it a coincidence that politics, housing and architecture alike were transformed in this period; on the contrary, the three were closely entwined. Housing was one of the main planks of social democratic politics (which in Britain meant Labour) and, in response, one of the areas in which anti-Labour political parties also wanted to make a mark. Architects saw in the advent of social democracy, with its state-funded programmes, both the opportunity and the necessity for a new kind of architecture, both as symbol and midwife of the new society emerging from the old. And it was through their claim to expertise in the design of housing for the working class – never before seen as a major area of architectural endeavour – that architects staked their claim to a leading role in the social democratic pageant.

Nor, equally, did the changes to our triad take place in isolation from the other changes – economic, technological, cultural – that were transforming the developed world at this time. The 'factory system', first identified as a phenomenon in Britain in the early nineteenth century, had matured and spread across the globe, generating not just the organised labour movement that was to provide the basis for social democratic politics but also the intense commercial competition between industrialised countries, notably Britain and its latter-day economic rival, Germany, that was to culminate in the outbreak of hostilities in 1914. But the kind of things that were being made, and the way that they

1

Fig. 1. Ford motor cars on the production line, c1912

were being made in factories, were also changing (Fig. 1). Thanks to its growing affluence, the working class was increasingly recognised as a consuming class; and fortunes were to be made on both sides of the Atlantic by those who, like Ford in America or Cadbury in Britain, perfected the methods of producing commodities for this new market, whether cars or chocolate. With these advances, many coming from the USA, there also arose new approaches to the organisation of production – scientific management, Taylorism, standardisation – which, at least until the USA lost its allure with the Wall Street crash in 1929, seemed to hold the key to improving efficiency and quality, and reducing costs. Nowhere was the appeal of this new approach greater than in relation to construction, with its hopelessly pre-scientific, 'rule of thumb' procedures and its chronic inability to deliver

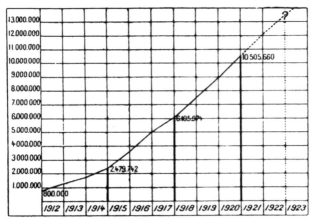

Fig. 2. Increase in automobile production in the USA since 1912 (from Le Corbusier, Urbanisme, 1925)

a decent home that the working class could afford (Fig. 2). For architects this suggested that the aesthetic ideal might lie not in the past (which to many now appeared quaint and outdated) but in the technology and organisational sophistication of the modern age. And, while the example of Henry Ford attracted attention worldwide, nowhere was it more potent than in a Germany re-equipping itself for modern life with the aid of American capital after the destruction of the German currency in the early 1920s. If Henry Ford could do it for cars (as it seemed he had, with the Model-T after 1912), why couldn't the same be done for housing?

At the epicentre of all these changes stood 'the war to end wars'. Developments that had been taking place before 1914 (the growing power of the labour movement over governments and the growing involvement of governments in the economic life of the country; the call for governments to promote or provide housing for the working class; the increasing reliance of governments on architectural experts for advice on housing provision and design) were accelerated a hundred-fold. In particular the working class was no longer needed just to make and consume things; now it was needed to bear arms and therefore had to be trained in military skills. The creation of a mass army (in Britain, five million men by 1918) transformed the relationship between government and working class: as the Home Office told the cabinet in London, 'in the event of rioting, for the first time in history the rioters will be better trained than the troops'[1]. In this context, working-class demands and grievances – not least over the shortage and inadequate standard of housing – assumed a new urgency which governments across Europe and even, albeit briefly, in the USA, (Fig. 3) sought to address[2].

Fig. 3. Kilham & Hopkins, Atlantic Heights emergency housing, Portsmouth, New Hampshire, USA, 1918

The model to which these governments overwhelmingly turned, was that developed in Britain in the decade or so before 1914 by the 'garden city movement', above all by Raymond Unwin. In an age when Fordism was entrancing the world – when every problem could be subjected to the scrutiny of 'the expert' – Unwin was, so far as housing was concerned, the expert's expert. Unwin had designed showpiece projects for each of the three strands that comprised the garden city movement – the 'industrial village' (Rowntree's New Earswick, York, 1902–), the 'garden city' (Letchworth Garden City, Hertfordshire, 1903–) and the 'garden suburb' (Hampstead Garden Suburb, London, 1905–); but it was with the garden suburb that he was most clearly associated and with which his theoretical work was primarily concerned. Whereas with the garden city Ebenezer Howard wanted to abandon the existing city and start from scratch in the agricultural countryside, Unwin's vision of the garden or satellite suburb started with the forces of suburbanisation that were already at work and sought to control them so as to produce a transformation in 'the dwellings and surroundings of the people' (Fig. 4).

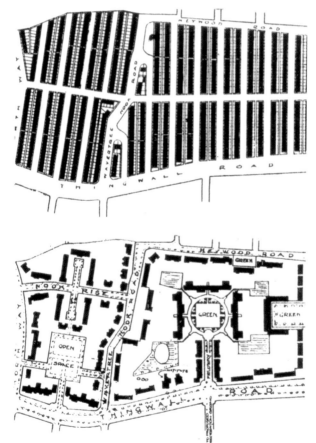

Fig. 4. Conventional versus garden city layout: Liverpool Garden Suburb, 1912, 'as it might have been' (top) and 'as it was'

It was of course precisely this transformative aspect of the Unwin model that, after the First World War, would make it so attractive to governments faced with a militarised working class: as one MP put it in 1919, by building houses 'on quite different lines' from those of the past, the government would prove that 'a new era for the working classes of this country' really had arrived[3].

Unwin set out his ideas in a series of cogently argued and lucid texts, starting with *The Art of Building a Home* (1901, with his partner Barry Parker) and *Cottage Plans and Common Sense* (1902), and taking in *Town Planning in Practice* (his magnum opus, generalising the lessons of Hampstead Garden Suburb, 1909) and *Nothing Gained by Overcrowding!* (1912). In Britain Unwin's legacy fed directly into the post-1918 housing programme; large parts of *Town Planning in Practice* were reproduced in the Tudor Walters Report of 1918 – the key government report, of which Unwin was the principal author – and his appointment as chief architect at the Ministry of Health consolidated his influence over the housing programme. But in other countries as well, the Unwin model provided the currency of post-war housing design, whether in France, where Henri Sellier planned a series of garden suburbs for Paris based on *Town Planning in Practice*, or in Germany, where a series of satellite housing developments was projected by the city architect of Frankfurt, Ernst May, one of Unwin's erstwhile assistants at Hampstead Garden Suburb.

But while by 1918 Unwin's work was viewed in largely technocratic terms, its origins and formation lay in very different ground. Unwin's thinking was shaped within the radical socialist movement of the 1880s and above all by his encounter with Edward Carpenter. Like Carpenter, the young Unwin looked to a millenarian transformation in society that was spiritual, ethical and communitarian all in one. In effect, although not in so many words, this was the arrival on earth of the New Jerusalem foreseen by Saint John in the Book of Revelation, which had inspired a lineage of English millenarians and visionaries from the middle ages through to William Blake. Carpenter was a transcendentalist poet (his hero was Walt Whitman) as well as a philosopher, socialist and campaigner who sought a new form of society and a new way of living. For Carpenter a major aspect of this emancipation was sexual – the freedom to love, and live openly with, another man. Although Unwin did not share this sexual orientation, Carpenter was his friend, mentor and hero and the main terms of Unwin's thinking were derived from Carpenter. Thus for Unwin commerce and convention were the enemies; what was needed was a rupture with convention that would both encourage and express a new way of living, in which a spirit of co-operation would replace competition and humanity would again live in harmony with nature. In terms of housing, this meant abandoning the conventional forms associated with the profit-based speculative builder and conceiving a new form of housing that would truly meet the spiritual and collective, as well as the individual, needs of the people who lived there. Such was the proposition, matured during Unwin's architectural partnership with Barry Parker from 1896, that was unveiled in print in *The Art of Building A Home* in 1901 and *Cottage Plans and Common Sense* in 1902 and on the ground, with the first houses at New Earswick in 1902–3 (Fig. 5).

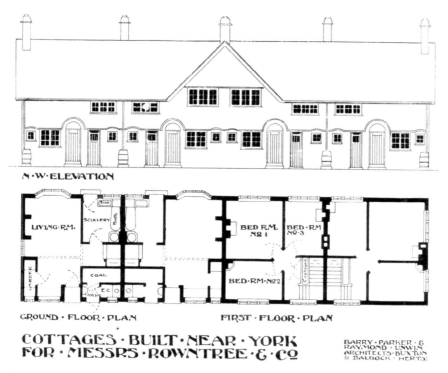

Fig. 5. Parker & Unwin, the first cottages at New Earswick, York, 1902-3

Unwin approached housing from the point of view of a Christian socialist who wanted both to improve the living conditions of the working class and to offer a route to the socialist future. But it was not necessary to agree with his politics to support his design precepts. For – and this was the key to the political import of housing in this period – housing was supremely ambiguous in its political affiliations. While the new housing offered improved conditions for the working class (and therefore attracted support from socialists) it did so in a way that left the main economic relations of capitalism untouched (and therefore attracted support from employers). Industrial capitalists like Lever, Cadbury and Rowntree saw in this new kind of housing an affirmation of the ability of capitalism to improve the lot of the worker; and it was the fact that the garden city movement attracted support from industrialists and landowners as well as radicals that made it so appealing to Lloyd George and the Liberal government in the years up to 1914. Similarly when, after the Armistice (November 1918) in 1918, Lloyd George, looking to the new kind of housing developed by Unwin to provide proof of the advent of the 'Land fit for Heroes', finally unveiled his housing campaign, opposition came not from industrialists (who were well aware of the adverse effects of the housing shortage on labour recruitment) but from two distinct sources. One was the *rentier* class (the owners of working class housing and their agents), whom the government could largely afford to ignore. The other, far more potent, was the City, whose financiers had arranged the advances to the government to pay for the war but did not see why they should do so to

Fig. 6. The largest municipal housing estate in the world: the London County Council's Becontree estate, commenced 1920

Fig. 7. Inter-war private enterprise housing in Romford, Essex

pay for the peace – and whose opposition in 1921 brought the housing programme to a premature halt. The advent of the first Labour government in 1924, however, re-activated the municipal housing programme and when the Conservatives – keenly aware of the need to offer social reform if they were not to lose out to Labour – returned to power, they decided to retain it, albeit with the proviso that the Building Research Station should seek a novel and cheaper form of construction for social housing. The outcome across Britain in the 1920s was the construction of low-density cottage estates designed according to (an inevitably pared-down version of) Unwin's precepts and often built of concrete or steel rather than brick (Fig. 6).

The local authority estates, however, were not the only ones built on Unwin's model. The great forces of free-market suburbanisation which Unwin had sought to reform proved themselves more amenable to change – albeit of course on their own terms – than he could have imagined or, in the form in which it turned out, than he would have wished. Speculative builders, whom Unwin in the 1900s had berated for building an unacceptable form of housing, were also persuaded, if not by the arguments, at least by the authority of the Tudor Walters Report and adopted Unwin's low density method, albeit with the design subtleties, as well as the social programme, largely omitted (Fig. 7). The upshot was that the kind of low-density perimeter block layout that Unwin had pioneered for the pre-war garden city movement as a radical break with convention had itself become, by the 1930s, the new convention – and indeed one that was to spread internationally, to become the standard form of housing in the Anglophone world[4]. The New Jerusalem that had been sought with such conviction by the young Unwin had, in a sense, arrived; although to most observers (including Unwin himself) it was not the New Jerusalem at all, but merely the old Jerusalem in a different guise.

Whether it indicates perseverance or merely lack of imagination, these are issues with which I have been engaged, on and off, over three decades. Within that period four more or less distinct phases can be identified in each of which the focus has differed, although the approach and method has remained consistent. While the essays which follow, written for a variety of audiences and for a variety of occasions, could have been grouped thematically or ordered by subject, on the whole it seemed more straightforward and transparent to present them simply in the order in which they were written.

The first phase was that in which I was engaged on my PhD, 'Homes fit for Heroes', which I commenced in the mid-1970s. The starting point for this was a curiosity about what my supervisor, Reyner Banham, would have called 'actual monuments'[5]. As items in the landscape, the municipal cottage estates of the inter-war years were sufficiently distinctive to be almost immediately recognisable: but what were the architectural ambitions and political motivations that lay behind them? These questions had a particular resonance in the troubled decade of the 1970s, as Britain entered a period of profound economic and social instability and the role of the state in maintaining the economic order came under intense scrutiny, particularly from those on the left.

I was fortunate in that the release of the relevant official papers under the 'Fifty Year Rule' meant that, for the first time, it was possible to answer these questions with some degree of certainty. I was also pleased to find another researcher, Simon Pepper from the University of Liverpool, who shared some of these interests, particularly in the wartime housing schemes that preceded the wider municipal programme. Therefore alongside the work for my PhD, published as *Homes fit for Heroes* in 1981[6], I collaborated with Simon Pepper on a pair of articles looking at this wartime programme (in which, inevitably, Unwin was one of the principal figures) and at the architectural debates, particularly over standardisation and mass production, that it helped provoke.

At this time I was teaching the architectural history programme at the Bartlett (University College London) with Adrian Forty and in 1981 we introduced a Masters course in history of architecture – the first of its kind in the UK. Part of that course was devoted to the architecture of social housing in Europe in the 1920s and this led me to look more closely at developments in France, Germany, Austria and elsewhere. A major theme that soon emerged was the way in which Unwin's precepts were taken up by policy makers and architects in Europe and the way in which, particularly in Germany and central Europe, they were modified in the light of the prevailing enthusiasm for Fordism and scientific management.

Funding from the Royal Institute of British Architects (RIBA) enabled me to collaborate with the American scholar Christiane Crasemann Collins in identifying for translation some key German-language texts charting this process, including that written by Karel Teige for the 1930 CIAM congress in Brussels.

The next phase centred on a study I started in 1983 of the socialist tradition in English architectural thought – Ruskin, William Morris, Philip Webb, Lethaby, Unwin, Penty and others – which was published in 1989 as *Artisans and Architects*[7]. Unwin often cited Morris, and behind him Ruskin, as influences on his development; and in the 1980s Morris was still widely regarded in architectural circles as a socialist exemplar. I was curious about this tradition: where it came from, what it was concerned with (and what it ignored) and what its relationship was to socialist politics and Marxism. Part of this study looked at the intellectual formation of Unwin who, it turned out, was far less influenced by this tradition, and far more by the liberationist philosophy of Edward Carpenter, than had previously been thought. Here was the explanation for many of the key themes in Unwin's thinking. A subsequent conference in Venice held to mark the centenary of *Der Städtebau*, the celebrated treatise on city planning published by the Viennese theorist Camillo Sitte in 1889, provided the opportunity to explore Unwin's debt to this major work.

By this stage I had left academia for publishing. In 1989 with Ian Latham I set up a new professional magazine, *Architecture Today*, and for some time this allowed little opportunity for historical research. In 2000 we started a sister publication *EcoTech*, a professional journal devoted to sustainable design. But I was intrigued by the historical antecedents of the issues that we were dealing with on a day-to-day basis; and this gave rise to the fourth phase, focusing particularly on alternative methods of construction. For the past decade or more advocates of sustainability have been demanding the adoption of constructional methods that are less damaging to the planet, and in this regard rammed earth (which requires very little energy to produce) is virtually unmatched. The 'bible' of rammed earth had been published in 1919 by Clough Williams-Ellis as part of a campaign to get the British government to use the technology for its post-war settlement programme and the resultant cottages at Amesbury, built in 1920, still form the benchmark for rammed earth construction in the UK. What had led the revivalists of rammed earth during the First World War to espouse this technology – and how much did it have in common with the objectives of the rammed earth revivalists of the present day?

Meanwhile, following the 1997 election, the Labour government was heavily promoting Modern Methods of Construction (MMC), also known as 'volumetrics' or 'modular systems' (a new version of the application to housebuilding of techniques from automobile production), all as though the previous episodes of 'prefabrication', 'industrialisation' and 'system building' extending from the 1920s to the 1970s had never happened. In an eerie re-run of the early twentieth century, the Blair government even commissioned a report (*Rethinking Construction*, 1998) from a committee chaired by Sir John Egan, the former head of Jaguar Cars, which was by then part of the Ford Motor Company. Just what was it with social democratic governments that made the pursuit of new methods of construction such an article of faith? The standard literature told us that it originated in the First World War, with Britain in the vanguard; the request from Unwin and the Tudor Walters Committee for research into alternative methods of construction in 1917 was cited as the start of a process that led to the opening of the Building Research Station in 1921 and in 1925 to its transfer to Garston where, re-named the Building Research Establishment (BRE), it has remained ever since. But the existing accounts were written almost entirely from a technical point of view. They therefore omitted what, for me, was the really interesting question – namely what led social democratic, and proto-social democratic, governments to identify constructional innovation as the key to their social programmes and to devote such efforts to achieving it. In the end this curiosity generated two essays (in both of which Unwin was again a leading figure), one on the origins of the Building Research Station in 1917–1921 and the other on its dramatic expansion at the behest of Neville Chamberlain in 1925.

One of the characteristics of the housing revolution of the period 1900–1930 was the extent to which it was based on publications by architects like Raymond Unwin and Clough Williams-Ellis, official reports like the Tudor Walters Report and the reports of the Building Research Board. It was through documents such as these that the new theory and practice of social housing were disseminated, nationally and internationally. But while some of these publications are held by national and specialist libraries, many are available only in London, in the official publications collection of the British Library or the National Archives at Kew. To make them more widely accessible, an accompanying CD has been included with this book, containing scans of some of the most important. These fall into four main groups. First, pre-war writings by Raymond Unwin on housing and town planning, including *Cottage Plans and Common Sense* and *Town Planning in Practice*. Second, British government reports on housing design and construction from the period 1913–18, including the report of the Women's Housing Sub-Committee and the Tudor Walters Report. Third, the key texts of the rammed earth revival, including the Clough Williams-Ellis book and official reports on Amesbury and other investigations. Fourth, texts relating to building research and the adoption of new methods of construction for state housing. This last group includes not only reports by the Ministry of Health and the Building Research Board but also the unpublished 1965 history of the Building Research Station by RB White, which hitherto has remained within the archives of the Building Research Establishment.

Thanks go to the many individuals who, in different ways over so many years, helped with the researching and writing of the essays, including George Atkinson, Charlotte Benton, Tim Benton, Iain Borden, Lesley Brooks, Nicholas Bullock, Jean-Louis Cohen, Christiane Crasemann Collins, the late George R Collins, Murray Fraser, the late Robert Gutman, Harry Harrison, Rowland Keable, Mervyn Miller, Gordon Pearson, Simon Pepper, James Read, FML Thompson, Robert Thorne and especially Robert Sinfield. Particular thanks also go to Simon Pepper and Christiane Crasemann Collins, the co-authors respectively of chapters one and two (Simon Pepper) and chapter six (Christiane Crasemann Collins), for consenting to the re-issue of these essays. Above all I am indebted to Adrian Forty, who over a period of 30 years has read and commented on these essays in draft form, and also to my publisher Nick Clarke and editor Sheila Swan of IHS BRE Press for their unfailing enthusiasm and support in producing this volume.

Thanks are also due to the original publishers for permission to reprint the essays, as follows: S Pepper and M Swenarton, 'Home front: garden suburbs for munition workers 1915–18' *Architectural Review* clxiii (June 1978) pp366–75; S Pepper and M Swenarton, 'Neo-Georgian maison-type' *Architectural Review* clxviii (August 1980) pp87–92; M Swenarton, 'An "insurance against revolution": ideological objectives of the provision and design of public housing in Britain after the First World War' *Historical Research* (formerly *Bulletin of the Institute of Historical Research*) liv no 130 (1981) pp86–101; M Swenarton, 'Rationality and rationalism: the theory and practice of site planning in modern architecture 1905–1930' *AA Files* iv (1983) pp49–59; M Swenarton, 'Sellier and Unwin' *Planning History Bulletin* 7 no 2 (1985) pp50–7; CC Collins and M Swenarton, 'CIAM, Teige and the housing problem in the 1920s' *Habitat International* 11 no 3 (1987) pp153–59; M Swenarton, 'Raymond Unwin: the education of an urbanist', in M Swenarton, *Artisans and Architects: the Ruskinian tradition in architectural thought* (Macmillan 1989); M Swenarton, 'Sitte, Unwin e il movimento per la città giardino in Gran Bretagna', in G Zucconi (ed), *Camillo Sitte e i suoi Interpreti* (FrancoAngeli, 1992) pp229–35; M Swenarton, 'The rammed earth revival: technological innovation and government policy 1905–1925' *Construction History* 19 (2003) pp107–26; M Swenarton, 'Breeze blocks and Bolshevism: housing policy and the origins of the Building Research Station 1917–21' *Construction History* 21 (2005–6), pp69–80; M Swenarton, 'Houses of paper and brown cardboard: Neville Chamberlain and the establishment of the Building Research Station at Garston in 1925' *Planning Perspectives* 22 no 3 (July 2007) pp257–81.

Chapter one

Home front

Simon Pepper and Mark Swenarton

The Housing and Town Planning Act of 1919 is generally considered to be the major turning point in the history of public housing in this country. Before the First World War the housebuilding activities of local authorities had been limited by the fact that any loss had to be borne by the rates; accordingly, the scope of those activities remained relatively small, amounting to no more than five per cent of the housing built in any year. The Act of 1919 marked a break with previous policy by transforming the optional power of local authorities to provide housing into a duty, and by providing a Treasury grant to absorb any losses in the Housing Revenue Account in excess of those that could be met by a local rate of a penny in the pound. The Act was conceived as a temporary measure to deal with the dual problem of the acute housing shortage and inflated costs brought about by the war, not to provide a general rent subsidy. Once the shortage created by the war had been met, the government hoped to be able to withdraw from the housing arena, leaving the processes of the free market to take their course. In the event, neither of these hopes was fulfilled. The 1919 programme was cut short prematurely when the spending cuts of 1921 limited building under the Act to 176,000 houses instead of the 500,000 originally contemplated. But successive governments were obliged to provide further subsidies, with the result that of the four million houses built between the wars, just under 30 per cent were provided by local authorities.

Despite the much greater numerical achievements of these later inter-war programmes, the 1919–1921 housing drive has retained its identity as what Marian Bowley called the 'first experiment' in direct state intervention[1]. It was, however, preceded by a programme of state-subsidised housebuilding that had arisen even more directly from the needs of the war. In the years 1915–18 the government found itself compelled, as part of its efforts to expand the production of munitions of war, to undertake the provision of accommodation for large numbers of workers in strategic industry. Not only temporary housing but also permanent garden suburbs and, in some cases, new townships were erected (Fig. 8). Although ignored by both architectural and social historians, it was this war-time housing programme that provided basic experience for the initiative of 1919[2]. The subsidies developed for munitions housing provided a model for the subsidy of the 1919 Act. Dr Christopher Addison who, as minister of health in Lloyd George's post-war government piloted the 1919 Bill through Parliament, gained his formative experiences of housing in the Ministry of Munitions. Most

Fig. 8. Frank Baines and others, Well Hall estate, Woolwich, 1915

significant for our purposes, the munitions programme provided the context for important changes in the architecture of public housing.

Most pre-war local authority cottage estates borrowed extensively from the picturesque quasi-vernacular of the garden city movement, with its steeply pitched gabled roofs, broken by dormers and elaborate chimney stacks. A few such schemes would still be built in the early 1920s, but most inter-war council housing was designed on quite different lines with uncluttered, medium-pitched roofs, restrained decoration and a minimum of projections from the simple, box-like shapes. The simplified inter-war municipal style is often regarded as the inevitable consequence of state sponsorship and, in so far as the houses were subsidised and controlled architecturally to an unprecedented degree, such a view is justified. The advent of the simplified style was, however, no accident rising out of the inadequacies of bureaucrat designers. It was formulated during the war by a distinguished group of architects, many of whom were leading figures in the garden city movement, some of whom remained in government service to exercise a decisive influence over the direction of post-war council house design.

The war-time housing problem and the Ministry of Munitions

The Ministry of Munitions may seem an unlikely source of inspiration for social experiments of this kind. The supply of arms and ammunition, however, was obviously of critical importance to the war effort and it was failure in this area, in particular the scandal caused by reports of shell shortages on the Western Front, which played a major part in the fall of Asquith's government in the summer of 1915. With the formation of the coalition, Lloyd George moved from the Exchequer to the newly formed Ministry of Munitions where he was joined by Christopher Addison as his junior minister. In his efforts to boost productivity and recruitment, Lloyd George imposed a number of controls which, although falling short of industrial conscription, severely limited the rights of munitions workers. It is for these measures that his Ministry is chiefly remembered[3]. Strikes and stoppages were made illegal, the unions were compelled to accept dilution of skilled trades, and workers were effectively tied to their places of employment by the system of 'leaving certificates'. Lloyd George himself described the Ministry as 'from first to last a businessman's organisation' and, up to a point, it was good business to care for the welfare of the workers[4]. Seebohm Rowntree, the cocoa magnate and industrial philanthropist, was appointed chairman of the welfare committee and used the post as a base from which to lobby for radical housing reform. The Ministry's housing programme, however, was regarded less as a welfare measure than as a compelling strategic necessity.

For a number of years before the war housebuilding had been in decline. Against this background the rapid expansion of war-time industry led to local housing difficulties serious enough to threaten recruitment and output and, in some places, to provoke civil disorder. The disturbances on the 'Red Clyde' are the best known product of a domestic crisis exacerbated by war[5]. Other smaller centres, however, experienced even greater pressures. In August 1914 Barrow-in-Furness contained a population of 70,000 which swelled to 75,368 by the end of the year, to 79,206 by the close of 1915 and to 85,179 by the end of 1916. The Royal Arsenal at Woolwich employed 10,866 on the outbreak of war: by 1917 the same factory complex employed 74,467. Still greater difficulties were caused by the need to locate explosives factories in relatively deserted areas. The cordite factory planned for Gretna would employ between 10,000 and 15,000 workers in a district estimated to be capable of accommodating only 4500 within a radius of 25 miles [40.2 kilometres][6].

At first temporary housing seemed the best solution; at least it did so to the Treasury, which insisted that all housing be temporary unless it could be shown either that there would be an obvious post-war demand, or that houses were needed to attract skilled workers, charge hands and supervisory staff (who, for the most part, would be married men with families)[7]. But the earliest hutted camps made heavy demands on imported timber which quickly became scarce. The need to conserve timber gave rise to a number of new concrete systems, but these cost almost as much as conventional permanent houses, and by the middle years of the war more permanent than temporary units were being built.

The construction of permanent housing schemes raised a number of wider policy problems which are illustrated by our earlier examples of Barrow, Woolwich and Gretna. Barrow had been a stable community enjoying above-average incomes as a result of the high wages paid by Vickers Shipbuilding, the firm which dominated the economic life of the town. Vickers had built a company village on Walney Island, but the rest of the town's housing was overwhelmingly owner-occupied – an unusual state of affairs for the time. Although the influx of war workers created a very real housing demand, plans to initiate a large-scale permanent housing programme were inhibited by fears of a post-war population slump, a housing surplus and consequent loss of value in existing property[8]. Something of this kind had already happened in Woolwich following the South African War (1899–1902)[9]. By 1914 however, Woolwich was on the growing south-east London suburban frontier and there was not much opposition to public initiative. At Gretna there was no alternative to public housing and, at the same time, no possibility of sustained post-war demand.

In different ways the three examples all made the case for subsidised housing. Certainly no private developer could be expected to invest in building at Barrow or Gretna, and local authorities would be reluctant to do so unless protected both against the current high cost of construction and what seemed to be a certain future fall in house prices and rents. To meet these difficulties, a number of subsidies were devised to encourage building by local authorities and private firms. Such systems provided either a capital grant to bridge the difference between pre-war and current building costs, or loans at low interest to cover current costs. Under the loan system, the houses would be valued three to five years after the war (by which time, it was confidently expected that prices would have returned to 'normal') and only the valuation price would be repayable[10]. It was in the Ministry-built schemes, where the government paid the full cost, that the question of rents became a matter of public policy.

In November 1915 the government had been forced to impose a rents freeze on existing houses, but this control did not apply to houses completed later in the war. What rents should be charged for them? If an economic rent was based on high war-time building costs, the houses would prove difficult to let. Subsidised rents, on the other hand, ran counter to the strongly held belief that public housing – like any other investment – should yield a modest return on capital. Yet faced by the necessity to attract skilled workers, rents were lowered to a point where practically no profit was made and, under pressure from tenants' organisations (pressures which included rent strikes), were pushed below economic levels.

In urban areas where building costs were low, rents went only a small way below the break-even point. Factories in deserted areas (and this applied with particular force to the Explosives Department's plants) needed to attract skilled family men away from populated districts. Here accommodation had to be provided and let at a loss and, after a series of ad hoc decisions, rents tended to be calculated at about 70 per cent of the economic level[11]. The 30 per cent subsidy was regarded by the Treasury as a special supplement to compensate the Ministry for unusually high costs; and it was this notion of compensation for extraordinary costs which, once established, was incorporated into the first peace-time subsidy of Addison's Act.

The housebuilding efforts sponsored by the Ministry of Munitions were both impressive and wide-ranging. Altogether the Ministry provided temporary hostels for 20,800 workers (most of them for women) and built 2800 temporary cottages, the largest concentration of these being at Elizabethville, a colony for Belgian refugees at Birtley, near Newcastle-upon-Tyne. In the long term, however, its most important achievement was the construction of 10,284 permanent houses on 38 different estates throughout the country. In less than four years the Ministry built slightly more permanent houses than had been constructed by the London County Council (LCC) and the Metropolitan Board of Works over the previous third of a century – and here it should be noted that the LCC was by far the largest and most active housing authority in the country[12].

Design work was handled by a number of different teams. The Department of Explosives Supply (DES) of the Ministry of Munitions set up its own design team headed by Raymond Unwin; HM Office of Works provided groups led by its principal architects, Frank Baines and RJ Allison; George Pepler headed another from the Local Government Board (LGB); HE Farmer came from private practice to be chief architect to the controller general of merchant shipbuilding. A number of private architects acted as consultants, among whom should be mentioned Gordon Allen, William Dunn and W Curtis Green, and Stanley Adshead, Stanley Ramsey and Patrick Abercrombie[13]. Although they included none of the leading local authority architects, it represented a surprisingly high proportion of the more experienced and innovative low-cost housing designers, and it was the different approaches adopted by these groups that gave the 1915–18 programme much of its architectural interest.

Well Hall

The first scheme was firmly rooted in the tradition of the picturesque. This was the garden suburb of Well Hall at Woolwich, built in 1915 to provide accommodation for the enlarged workforce at the nearby Royal Arsenal. Although carried out at great speed – the entire estate of 1000 houses and 200 flats being conceived, designed and built during 1915 – the Well Hall scheme was nonetheless a tour de force of picturesque design, in which the components of the 'old English village' were assembled with a virtuosity exceeding anything attempted by Unwin, even at Hampstead Garden Suburb. Diversity of materials and external finishes (timber-framing, tile-hanging, slate-hanging, stone, brick and rendering in various degrees of roughness) was matched by a complexity of shape and silhouette (gables, dormers, overhangs, tunnels and various other projections and recessions) to produce an architectural ensemble that seemed centuries apart from an age of 'total war'.

To some extent this diversity was a response to the difficult materials supply situation: the architects used whatever they could lay their hands on in an effort to maintain the breakneck speed of the project. But there can be no doubt about their intention to exploit to the full its architectural potential. A number of the original drawings which are preserved in the Woolwich local history collection shows the composition of entire street elevations, both front and back, complete with colour wash and shadow projections and a host of minor adjustments to the standard terrace plans.

The layout of the estate (Figs 9 and 10) followed many of the precedents for low-density development established by Parker & Unwin and their colleagues before the war. Large blocks were created, with the houses distributed along the perimeters and the centres of the blocks occupied by generous gardens[14]. On the more steeply sloping parts of the site to the east of Well Hall Road, the

Fig. 9. Frank Baines and others, Well Hall estate, Woolwich, 1915, layout, phase one (above) and Fig. 10. phase two (below)

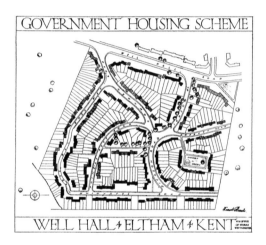

site roads followed the contours and are relatively closely spaced. The second phase of the scheme, to the west of the main road, was on more level ground. Here the blocks are much wider and the backlands have been exploited by an internal quadrangle served by narrow roads, built to a reduced specification, similar to those pioneered by Unwin at Hampstead Garden Suburb and known as 'carriage streets' in his major theoretical work, *Town Planning in Practice* (1909) [CD]. Site roads and buildings both demonstrate a preference for the picturesque. Following the view of Camillo Sitte – or rather, that of his translator Camille Martin – that 'the ideal street must form a completely enclosed unit'[15], the houses at Well Hall were assembled in terraces of up to 16 units; but with the careful staggering of recessions and projections, and the winding streets, the association is clearly with the medieval or early-modern town rather than the contemporary terraces of the speculative builder (Fig. 11).

Well Hall was designed by the Office of Works architect, Frank Baines, who continued to explore the picturesque idiom at other munitions housing schemes,

Fig. 11. Frank Baines and others, Well Hall estate, Woolwich, 1915, typical street

most notably the Roe Green Village, near the aircraft factories at Hendon. This was a smaller and somewhat less exuberant essay in vernacular design, in which the proportion of cottage flats (that is, flats in two-storey blocks) was increased to nearly 50 per cent. Like Well Hall, it is now a conservation area.

Reviewing Roe Green in January 1918, *The Builder* claimed to detect a schism in the architectural world between the supporters of the picturesque and the followers of Neo-Georgian, and concluded that 'the advocates of what we may term the picturesque traditional are at present in the ascendant ... '[16]. Time was to disprove this assertion. Whilst the use of a variety of materials provided one answer to the problem of war-time shortages, there was a growing school of thought which looked to the standardisation and mass-production of materials and fittings to overcome this difficulty in both war and peace. The design of the cottage, it was argued, should be simplified as far as possible. By the end of the war this had become the official orthodoxy, accepted alike by government and the great majority of housing 'experts'[17]. That this was so was largely due to the influence and activities of Raymond Unwin, in particular to the practical demonstration of simplified design afforded by the munitions villages built under his aegis. The most important of these was the development around the explosives factory at Gretna.

Gretna

In June 1915 it was decided by the then newly-established Ministry of Munitions to build a huge explosives factory on a remote site between the small villages of Gretna and Eastriggs, in Dumfries (Figs 12 and 13). The site offered excellent rail and road connections, could be served by coastal barges and, being on the west coast, was safe from German naval bombardment. Since accommodation would have to be provided for most of the construction and factory workforce,

Fig. 12. Raymond Unwin and others, Gretna, Dumfries, 1915, layout

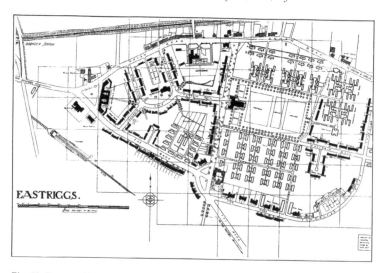

Fig. 13. Raymond Unwin and others, Eastriggs, Dumfries, 1915, layout

a housing branch was set up by the DES. Raymond Unwin was seconded from his post of town planning inspector at the LGB to direct the project, assisted by SB Russell and two former colleagues from his Hampstead Garden Suburb days, Geoffrey Lucas and Courtney Crickmer.

Following initial Treasury directives, the housing was to have been almost entirely in temporary wooden huts. Timber shortages, however, soon forced the architects to introduce a number of concrete units and, in May 1916, Addison authorised the construction of permanent housing[18]. The cost factors underlying this decision were set out in a minute prepared by GH Duckworth of the Ministry of Munitions following a visit to Gretna in February 1916[19]. At that stage £218,000 had been budgeted for factory workers' huts, and £40,000 for the construction labourers'. These sums would house about 9500 workers, on the then current costing of £27 per head for dormitory-style hutted accommodation. Hospitals and stores would cost £5700; shops and schools £26,000; and site development (roads, drains and main services) another £108,000. It was the high proportion of the total bill represented by these overheads (in particular the development costs) which persuaded the Ministry to invest in permanent buildings. By the end of the war 1038 permanent cottages had been built, including 97 units which had been obtained from the conversion of brick-built 'cottage shells'. This housing, together with a handful of surviving 'temporary' huts, forms a large part of Gretna's present stock.

Design and construction proceeded at a furious pace. Following the decision of June 1915, work on drains, roads and foundations started in August and by October 5000 construction workers were on site. The end of the year saw between 8000 and 9000 labourers commuting on special trains from Carlisle, which had become the bridgehead for the operation and which, among other sacrifices, endured government control of its pubs to minimise drunkenness among the horde of navvies[20]. By the end of 1917 the township had reached its maximum combined population of 24,000 construction and factory workers, some of the former staying on to man the factory. Well Hall had been criticised by Sir William Lever (builder of Port Sunlight) as a mere 'suburb' because it lacked communal facilities[21]. Gretna, by contrast, was more a small town than a large estate, its population being served by a dozen shops, bakery, laundry, central kitchens, post office, cinema, hall, dental clinic, schools and institute. There were no pubs, but spiritual refreshment was provided by no fewer than five large new churches. The architecture of these followed denominational lines; the two Espiscopalian churches being Gothic, the two Catholic churches in brick Italian Renaissance, and the Presbyterian church in Italian style with rough-cast walls and a red tiled roof.

The Gretna housing was the first major demonstration of the theory of simpli-fication in design (Figs 14 and 15)[22]. Gone were the broken roof-lines, complex shapes and variety of materials of Well Hall and Roe Green; Unwin's team used simple hipped roofs, eliminated dormers and projections, and restricted external finishes to fair-faced brick or rough-cast blockwork. Some of the cheaper units were little more than brick boxes, others – particularly the public buildings – boasted Neo-Georgian detailing. In general, the housing relied for its visual effect on what an earlier departmental committee in 1913 had called 'breadth and simplicity of design ... orderly arrangement of the buildings, the observance of good proportion ... and other elements of good design'[23] [CD].

The permanent housing at Gretna and Eastriggs was concentrated around the two village centres and distributed along the main site roads where, to some extent, it concealed the temporary hostels which were laid out in a rather military grid. But the 'orderly arrangement' of Gretna's permanent housing meant something quite different from the sinuous streetscape of Well Hall. The Gretna houses were either semi-detached or grouped in short terraces of four to six units. Where Baines had achieved more or less continuous frontages, Unwin – who was also conscious of the need for continuity – contrived this effect by placing brick outhouses and screen walls between otherwise isolated blocks. As the Tudor Walters Report [CD] was to put it, 'these links ... tend to remove the objectionable appearance of the repeated gaps'[24]. The layout devices employed by Unwin's team were the cul-de-sac, which exploited the backland in a low-density block, and the close or three-sided square, which was set back from the main road to save development costs. All of these devices – links, culs-de-sacs

Fig. 14. Raymond Unwin and others, Gretna, Dumfries, parlour houses (above) and Fig. 15. hostels (below)

and closes – were repeated in the other major scheme built by the DES at Mancot, between Chester and Queensferry. In the years that followed, the brick-built links tended to be omitted from all but the most lavish schemes, but the cul-de-sac and the close became characteristic features of inter-war site planning[25].

New methods of building

At Gretna timber shortages and high development costs had led to the decision to provide permanent buildings. Elsewhere, the shortage of both timber and brick led to experiments with new methods of building, experiments which gave rise – at Unwin's instigation – to the foundation in 1917 of the Building Materials Research Committee, the forerunner of the Building Research Establishment (see chapter ten)[26]. Since steel was at a premium for arms' production, the obvious alternative was concrete blockwork; and a number of munitions villages aroused particular interest for their use of the patent winget system of site-cast concrete blocks, a system which had been introduced a few years before the war.

The Crayford Garden Village housed workers at the Vickers-Maxim machine gun factory and was designed by Gordon Allen, consultant architect to the Rural Housing Organisation Society (one of a number of non-profit making societies which offered technical assistance to enlightened landowners, railway companies and collieries in providing housing for employees). As in the other schemes, there was a sharp distinction between the classes of accommodation provided: four-bedroom houses for supervisory staff, three-bedroom parlour houses with kitchen-dining room and a front parlour for charge hands and foremen, two- or three-bedroom non-parlour types for ordinary skilled workers. More than half of the 600 houses in the village fell into the latter category and it was these that were built out of winget blocks, rough-cast and colour washed. For many years Allen had interested himself in the design of low-cost housing and, at Crayford, seems to have built the cheapest schemes of the war-time programme. A small house at Well Hall had cost £465 in 1915, the average cost for the scheme being £622. These prices were well above initial estimates, the excess being attributed to the use of a prime-cost contract and the huge overtime and weekend working payments. In 1916, three-bedroom non-parlour types cost £350 at Gretna, while parlour types cost £380 (excluding site development charges). Gordon Allen claimed to have built three-bedroom parlour types at Crayford for less than £200, and to have achieved an average unit cost of £325, including roads, sewerage, lighting and water supply. Although only the Crayford scheme compared favourably with pre-war costs, all of the war-time projects were well below immediate post-war prices which sometimes reached as much as £1000 per unit.

Similar construction methods were employed on two well-known estates at Chepstow, where housebuilding was being sponsored by the controller general of merchant shipbuilding. HE Farmer was in overall control of what was to have been a massive housing development on both banks of the Wye. His scheme for the Bulwarks Village was completed shortly after the war and received praise for the quiet dignity of its simplified rendered-block house-types. The nearby Hardwick estate, designed by William Dunn and Curtis Green, was built in fair-faced blockwork with intermediate floors of hollow-pot concrete, again to save timber (Fig. 16). The Hardwick estate was completed in 1917, after which Curtis Green joined the forces. Dunn, a much older man and one of the pioneers of pre-cast concrete, kept the practice going and delivered a much publicised lecture on the scheme at the Royal Institute of British Architects (RIBA) just after the Armistice. The quality of the design (which fell architecturally between the picturesque and the simplified), plus William Dunn's enthusiastic advocacy of concrete blockwork as a housing material, did a great deal to create a favourable climate of opinion for non-traditional methods of construction after the war.

Fig. 16. William Dunn and William Curtis Green, Hardwick estate, Chepstow, 1917, concrete block cottages

Lessons for the design of post-war housing

By the early part of 1918 it was certain that the promised post-war public housing would be developed on garden suburb lines at the low densities favoured by housing reformers – 12 houses to the acre [0.4 hectare] in urban districts and eight to the acre [0.4 hectare] in rural areas[27]. It was by no means clear, however, which direction the architecture of public housing would take. The evidence available to *The Builder*, as we have seen, suggested that the picturesque school was in the ascendant, an impression confirmed by the prominence of picturesque designs in the prize list of the competition organised in 1917 by the LGB and the RIBA[28]. Although there was supposed to be a general ban on the publication of plans and photographs of munitions housing schemes, the censor's veto had been applied with such unevenness that, on the evidence of the few published schemes, the designs of the Office of Works seemed typical of munitions housing as a whole. Well Hall was reviewed as early as 1915 in the *Town Planning Review* and in 1916 in *The Architects' and Builders' Journal*. Roe Green was featured in *The Builder* at the beginning of 1918. Far from there being a total security clamp-down, Well Hall actually enjoyed extensive publicity abroad: it was reported that 'Propaganda work in neutral countries has been carried out by the showing of films taken of this great Government scheme ... '[29]. It was not until just before the Armistice that the work of Unwin's group at the DES was reported and illustrated in the British architectural press[30].

Until then, those who had made the journey to Gretna had not been over-impressed with its appearance. In February 1916, before the erection of the permanent buildings, GH Duckworth had reported to the Ministry of Munitions that 'the appearance of the township is at present forbidding'[31] and in 1918, a close associate of Unwin's, Mrs Alwyn Lloyd, felt that the 'whole effect of severely simply designed red brick and slate roofed houses is depressing'[32].

It was only during 1919 that the architectural journals took up the cause of the DES. Until then it had been Well Hall that had been taken as the model for post-war work. The comment of the *Building News* is typical: 'It is universally recognised that the standard set by HM Office of Works in this [Well Hall] and other schemes ... is ... a very valuable guide as to what should be done'[33]. During 1919, however, schemes by the DES and their architectural allies began to receive favourable attention and praise. In its review of George Pepler's designs for the housing at Coventry and Avonmouth, *The Architect* made explicit reference to the work of both the DES and the Office of Works: 'Like Mr Raymond Unwin's Gretna designs, the architectural quality of the work is at once simple and effective, and is to our mind more pleasing than the more elaborate work carried out by the Office of Works at Woolwich and at Hendon'[34]. In his book, Gordon Allen illustrated several cottages at Gretna and commented that 'a pleasant Georgian character is the external note'[35]. *The Architects' Journal* made similar points in its 1919 review of the housing built for Dorman Long at Redcar by Adshead and Ramsey, with Patrick Abercrombie as consultant:

> The elevations are almost severely plain, depending as they do for interest in their grouping, the careful disposition and proportion of the windows and the studied details of the doors.... The elements of these modest and charming Georgian buildings, so characteristic of many of the Yorkshire villages, lend themselves admirably to a system of standardisation inseparable from any modern housing scheme which is to be both effective and economical[36].

The first indication of the way official views were developing behind the scenes came with the publication of the Tudor Walters Report in November 1918. This body, known in housing circles as the 'experts' committee', had been set up by the president of the LGB when the government first announced its intentions for post-war housing in July 1917. Although often taken as a housing manual, the report was much more than this. It was the first (and in many ways still the best) thorough-going treatise on the theory and practice of low-cost housing design, encompassing matters as diverse as town and site planning, internal layout and servicing of houses, availability of building materials and new systems of construction. Experience gained in the munitions housing programme was brought directly to bear on these questions by the appointment of both Baines and Unwin to the committee. But Baines resigned only a month before the report was signed, and the finished product bore clearly the traces of Unwin's hand. Reproducing plans of Gretna cottages and diagrams from Unwin's earlier publications, the report argued unequivocally for the benefits of simplification: 'Simple, straightforward plans will usually prove most economical.... Ornament is usually out of place and necessarily costly both in first execution and in upkeep. The best effects can be obtained by good proportion in the mass and in the openings.'[37]

Some of the committee's conclusions went directly against the kind of picturesque design employed by Baines at Well Hall and Roe Green:

> An important economy can also be effected by a simplification in the design of the roof itself. It is very doubtful whether the introduction of dormers is conducive to economy and, though it may reduce the cube, it will probably increase the actual cost of the building. Generally, it is better to run the eaves in an unbroken line immediately over the first floor window heads and avoid the numerous expenses entailed in capital cost and maintenance when the roof is broken by dormers and flats[38].

Where there were divergences between the Office of Works and the DES (as, for instance, over the provision of flatted cottages, which Baines supported and Unwin opposed) the report came down clearly on the side of the latter. By thus articulating the views of the group associated with Unwin, the Tudor Walters Report gave those views the status of government policy.

Equally important for the future of public housing was the staffing of the government department responsible for the implementation of the new policy. The Office of Works made a bid for this role. In December 1918 the first commissioner of works, Sir Alfred Mond, proposed that supervision of local authorities should rest with his office which 'has not only the necessary technical staff but has throughout the war been engaged in the erection of very important Government Garden Cities, such as the Well Hall estate, Woolwich, which is now regarded as one of the models for this type of work ... '[39]. However, the task was given to the department traditionally responsible for housing, the LGB, now in its new and much expanded form as the Ministry of Health[40].

By the middle of 1919 the major appointments in the housing department of the Ministry of Health had been filled with men from the Ministry of Munitions, in particular with staff from the housing branch of the DES. The minister was Addison, whose only previous practical experience of housing had been at the Ministry of Munitions between 1915 and 1917. Sir James Carmichael, formerly chairman of the Munitions Works Board, became director general of housing. Two posts of chief architect were created – one in charge of housing layout, the other of house design – with powers of control over the form of the subsidised housing. Unwin got the first job; the second went to Russell, his former assistant at the DES[41]. Courtney Crickmer from the Gretna team served for a short time as deputy chief architect, and HE Farmer became one of the Ministry's regional commissioners for housing[42]. William Dunn retired from practice in 1919 (when his partner returned from France) to become the architectural member of the first housing advisory council and chairman of the standardisation and new methods of construction committee[43]. Many others obtained middle-rank posts, joined local authorities or launched their private practices with public housing commissions[44].

The extent to which the views of Unwin and his colleagues were identified with official policy became clear with the publication, in April 1919, of the *Manual on the Preparation of State-aided Housing Schemes*. It had previously been stated that the manual would contain the prize-winning designs from the LGB–RIBA cottage competition, but there was no sign of these schemes in the published document, which provided instead a précis of the Tudor Walters Report, with considerable emphasis on the merits of simplification. It would be wrong, however, to attribute the post-war success of simplification entirely to official propaganda and control. In part it was a response to economic and political circumstances. The scale of the post-war housing programme, together with the vastly inflated cost of building, made economy essential in the eyes of the Treasury. Yet at a time of political crisis, when an uprising of the working class led by the returning 'heroes' was daily feared by the government, it was clearly impolitic to attempt to hold down costs by the traditional method of lowering standards. Simplification appeared to offer a solution to the dilemma, a means of economising that was not politically contentious. It was indeed primarily as a means of saving money that simplification appealed to the government, and some architects were prepared to justify it on this basis alone. Ewart Culpin, for instance, writing in 1917 in the *Journal of the American Institute of Architects*: 'The logic of circumstances has forced upon cottage architects the conclusion that the simple type of cottage is the one which must now be concentrated upon ... Building has become, I fear, permanently dearer, and we must cut our coat according to our cloth ... '[45]. Before the end of the war, even the Office of Works had turned away from the excesses of Well Hall. The scheme designed by RJ Allison for the Royal Aircraft Factory at Farnborough in 1917–18 was largely composed of non-parlour houses, built on straightforward lines, with tile window cills, blockwork partitions, concrete floors and hipped roofs 'designed to avoid cutting as far as possible, thus reducing waste of material (timber) to a minimum'[46].

There were also more positive grounds for espousing simplicity of design. To arts and crafts designers of a certain persuasion it was to be preferred on aesthetic grounds as well. Faced with the harsh realities of low-cost housing, the picturesque and ornamental could be seen as a 'sham', a 'pretence' reminiscent of the speculative builder. To architects like Unwin, 'rightness of form' could be obtained only by a simple type of house without any false pretensions. This argument was expressed by Unwin as early as 1902 and was stated fully in 1906 by – ironically – Baines's former mentor, CR Ashbee: in designing a dwelling that the working-man could afford, the designer must 'eliminate all aesthetic super-fluities, all hand-work, and above all every pretence at it ... (making it) as simple, severe and economical as it can be made'[47]. This line of thought was reinforced by the rather different arguments emanating from the departments of architecture and civic design at Liverpool University. Here modernist and *Beaux-Arts* tenden-cies were combined with a hearty dislike for what Lionel Budden denounced as 'little coteries of romantic enthusiasts'. In April 1916 two important articles appeared in the *Town Planning Review* by Budden and Adshead who, addressing themselves to the problem of what they called 'lesser domestic architecture', rejected in toto the romantic tradition of garden suburb architecture. Instead they called for an officially-sponsored 'standard cottage' which would:

> depend for any attraction that it may possess, not upon the toolmarks of the workman, nor upon its peculiarity or idiosyncrasy, nor in a word upon its indi-viduality, but upon more general characteristics such as suitability to purpose and excellence of design. It will not be the home of an individual ... but the home of a member of a certain class of the community ... And thus with the well-designed standard cottage we shall get, not a village where each dwelling expresses the foibles of its tenant, or what in practice usually happens, the expression of these as understood by an architect; but instead a thoughtfully composed collection of similar units, each of which contributes to the making of a composite design, and each of which bears the impress of being part of a carefully considered scheme[48].

That the doctrine of simplification, along with the other precepts of Unwin and his colleagues, could be so wholeheartedly adopted in 1919 was due, in large measure, to the changed nature of the post-war housebuilding programme. Before the war the housing experience of the local authorities had been predomi-nantly in the field of slum-clearance and re-housing, involving the erection of tenement blocks on central urban sites. The policy of 1919, by contrast, sought to relieve the general housing shortage by providing accommodation of a higher standard for the more prosperous members of the working class, and to do so by building low-rise units, built to low densities on suburban estates. It was this change in policy that rendered previous official experience in design largely redundant and opened the door to a new group of architects who had both recent experience of building under emergency conditions and a well-resolved architectural philosophy for the municipal suburbs.

Chapter two

Neo-Georgian *maison-type*

Simon Pepper and Mark Swenarton

Modern architecture provides surprisingly few examples of buildings designed in accordance with a previously stated programme. There is, of course, no shortage of written statements. A preoccupation with debate has always been a prominent feature of modern architecture. However, so much of the polemic either yielded no recognisable architectural product or was obviously post-rationalised, that the subject of this study is something of a rarity: an architectural scheme which was faithfully constructed to a model described in a series of articles published shortly before its design.

The industrial village of Dormanstown was started in 1917 to the designs of Adshead, Ramsey & Abercrombie (Fig. 17). The ideas on which the design was based had been set out in a series of articles printed in the early volumes of the *Town Planning Review*, culminating in a group of papers in April 1916. Dormanstown and a number of other war-time schemes – notably the townships of Gretna and Eastriggs designed by Raymond Unwin and his colleagues at the Ministry of Munitions – later became familiar as prototypes for the large-scale state-subsidised housing schemes of the 1920s. The close relationship of the built scheme to the publications which preceded its design makes Dormanstown particularly interesting, for it allows us to examine the theoretical basis for a type of architecture that in Britain is not normally credited with anything so intellectually respectable as an underlying philosophy. As is well known, the design of standardised mass housing for the working classes was a central issue for the theoreticians of the European modern movement: Dormanstown and its literature provides an insight into a parallel debate conducted in Britain by men of a quite different architectural persuasion.

Fig. 17. Adshead, Ramsay & Abercrombie, Dormanstown, 1918

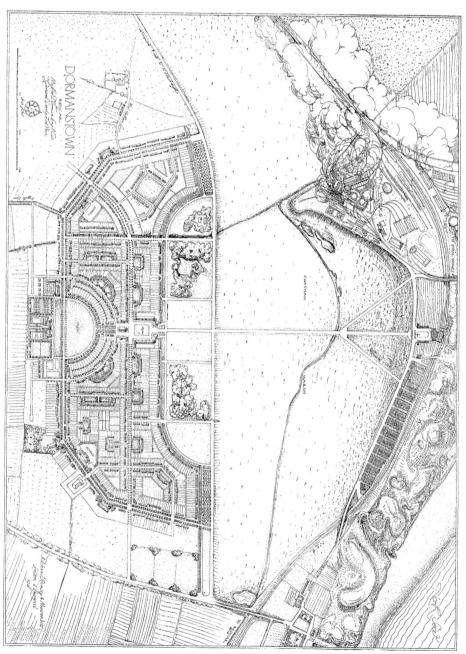

Fig. 18. Adshead,
Ramsay & Abercrombie,
Dormanstown, original
layout, 1918

Dormanstown

Dormanstown was built towards the end of the First World War to house workers at Dorman Long's iron and steelworks at Coatham. Today the scheme has been engulfed by the sprawl of industry and suburb between Redcar and Middlesbrough but, when it was planned to accommodate the greatly expanded war-time workforce, the development was isolated among open fields on a flat, windswept, coastal site. To its designers Dormanstown was a self-contained 'village', rather than an estate. Schools are located in the western and eastern halves of the scheme. To the south a community clubhouse overlooks the crescent 'green'. To the north is a 'market square' surrounded by shops and overlooked by a church. Four more church sites are indicated on the southern boundary road[1]. Similar facilities, of course, are to be found in other planned villages, although rarely on the scale of Dormanstown. What distinguishes Dormanstown from the best-known pre-war schemes – Hampstead Garden Suburb, Bournville, Port Sunlight – is the formality of its design.

The layout (see Fig. 18) was a spectacular example of the baroque planning that was to become popular – even conventional – between the wars. A central north-south road ran from the green, through the market square, to the factory. The rest of the village and its public buildings were disposed symmetrically about this axis. Where departures from absolute symmetry were necessary – usually to accommodate existing roads – they were so skilfully integrated into the plan that they became almost unnoticeable. All of this shows a very different approach to residential townscape from that made famous by Raymond Unwin in his early days. Unwin stood for the layout of streets based on topography and economy and drew from his reading of Camillo Sitte the notion of the street picture as the overriding concern of the planner (see chapter eight)[2]. To him the idea of a preconceived overall site geometry would have been an anathema.

*Fig. 19.
Adshead,
Ramsay &
Abercrombie,
Dormanstown,
detached and
semi-detached
houses, 1918*

Equally at variance with garden city orthodoxy was the design of the houses. Where at Hampstead Garden Suburb gables and dormers abounded and bob-and-string latches were never far away, at Dormanstown the buildings and their components are all assembled within the discipline of the Neo-Georgian style (Fig. 19). The houses themselves are simple brick or stuccoed boxes with low-pitched, hipped roofs. Wherever possible the chimneys are grouped into simple, central stacks. The use of six-panelled doors, pedimented porches, multi-paned windows, bays and moulded cornices combine to reinforce the impression of eighteenth-century development. At Hampstead Garden Suburb the intention had been to recreate the old English village. At Dormanstown we have a Georgian 'New Town'.

Fig. 20. Adshead, Ramsey & Abercrombie, Dorlonco steel-framed houses at Dormanstown under construction, 1919 and Fig. 21 (below) completed

So complete is the eighteenth-century idiom at Dormanstown that it comes as a surprise when one realises that many of the houses were built from a highly industrialised, non-traditional system. The first 300 units were built during 1917–18 in load-bearing brickwork; although even here steel joists and pre-cast concrete floor slabs were used to save timber, which during the war was even more scarce than steel. For the post-war units a steel-frame system was employed. It comprised a light steel frame of L-sections, bolted together on site to provide roof trusses and wall stanchions (Fig. 20). Expanded metal lathing was then fixed to the frame with wire ties and sprayed with a two-inch [5.1 centimetres] thickness of sand and cement rendering to form the outer skin of the walls. Lightweight concrete blocks were used for the inner leaf of the external walls and for internal partitions. The intermediate floor of the early post-war houses was cast in situ using expanded metal lathing as permanent shuttering and nominal reinforcement. The roof was finished in tiles or slates (Fig. 21)[3].

Not surprisingly perhaps, the thin external rendering was prone to cracking and, when this occurred, water would find its way into the cavity, corroding the steelwork and the wire holding the structure together. After serious problems had been encountered in the early 1920s, the system was redesigned with an outer skin of brick or rendered blocks. At a time when the building industry continued to suffer from shortages of skilled tradesmen, the system offered the possibility of employing steel workers and labourers in place of bricklayers and carpenters and, by getting the frame and roof erected early in the construction process, of carrying out the subsequent operations under cover[4]. For the steel manufacturers the system represented another outlet for their product when the demands for munitions fell away after the Armistice. The design was marketed as the Dorlonco system and, in its modified form, became one of the most successful non-traditional building methods of the inter-war years[5].

The Liverpool school

Before turning to the ideas behind the scheme, something should be said about its authors, most of whom were connected with the University of Liverpool. Stanley Adshead was the first Professor of Civic Design at Liverpool before resigning in 1914 to take up the new chair in Town Planning at University College London[6]. For some years before this move he had been in partnership with Stanley Ramsey[7], with whom he designed the housing scheme for the Duchy of Cornwall estate in Kennington, an early essay in Neo-Georgian working-class housing which was highly praised upon publication just before the start of the First World War (Fig. 22)[8]. Patrick Abercrombie was appointed to the staff of the Liverpool Department of Civic Design on its foundation in 1909 and succeeded Adshead in both the Liverpool and the London chairs[9]. He also edited the *Town Planning Review* which as the first and, until 1914, the only professional planning journal occupied a position of unique authority.

*Fig. 22. Adshead &
Ramsey, Duchy of
Cornwall estate,
Kennington, 1914.*

The man who appointed both Adshead and Abercrombie was Charles Reilly[10], Roscoe Professor of Architecture at Liverpool, an immensely dynamic and forceful man who not only built his department into the leading university school of architecture – arguably the only real rival to the Architectural Association between the wars – but stamped its curriculum with 'the straightforward, large scale classic, with all the implications of axial planning that followed.'[11] To Reilly, town planning was but 'an advanced form of architecture ... an expression of, almost one might say, the climax to the teaching of the School of Architecture.'[12]

Adshead, Abercrombie and Reilly were the central figures in what we have called the 'Liverpool school'. The term is useful, notwithstanding the proliferation of contemporary schools, groups and movements, since it denotes a close group of colleagues from a single institution, sharing a distinctive view of architecture and planning. In architecture they campaigned for monumental classicism, with such good effect that the style was later to be described by Randall Phillips as the 'Liverpool manner'[13]. In the field of planning, the school opposed what they saw as the anti-urban tendencies of the garden city movement, arguing for the reconstruction of city centres – on monumental lines, of course – rather than the direction of resources into suburbia or beyond. Some members of the Liverpool school went further: Lionel Budden, Liverpool graduate and lecturer and Reilly's eventual successor in the Roscoe chair, denounced garden suburbanites as 'little coteries of romantic enthusiasts'[14]. The Liverpool school were unashamed urbanists and classicists, dismissive of the provincial pretensions as well as the picturesque neo-vernacular of so many garden suburbs. As city planners they were also conscious of the pressing need for a large-scale housebuilding and reconstruction programme in the major conurbations. It was the First World War, however, which eventually created the necessary political conditions for a state housing programme and which, more immediately, provided an opportunity for the school to put into practice their ideas on mass housing design.

Production and society: the standard cottage

In April 1916 a major congress on post-war housing took place[15] and it was to coincide with this event that the *Town Planning Review* published three articles from the pens of Adshead, Ramsey and Budden[16]. Drawing upon suggestions already published by Reilly and Abercrombie[17], the authors of the three articles of 1916 presented the case for giving the post-war state-aided housing programme an architectural treatment very different from that associated with most garden suburbs. Like most of what passes for architectural theory, their presentation was stronger on assertion than argument, but running through their writings there can be detected three notions which, taken together, provided the rationale for the architecture to be adopted at Dormanstown. These notions involved, respectively, the nature of modern production, the nature of modern society and the role of taste.

As the 1916 congress made clear, the post-war programme would be on a different scale from anything conceived before 1914. In the garden city movement architects had been dealing with individual clients, or small groups of co-partnership tenants; in the post-war programme it would be mass housing for a mass clientele. For a scheme on this scale, stated the Liverpool school, the mass production of standardised components was essential. Using language and argument curiously similar to that employed by Muthesius, Gropius and other modern movement theorists, Lionel Budden declared that in mass housing individual treatment of the elements of the house was impossible:

> A standardisation of elements and details of design is the logical result. It is a consequence, whose inevitability becomes the more obvious the better its origins are appreciated. Stock patterns have been evolved by an irresistible demand, by the operation of economic laws to which no effective opposition is now possible, even if it were desirable. Stock windows, stock doors, stock fireplaces, have not been forced upon a reluctant market. They were not the product of amateur experiment, of caprice. They have been created in response to needs which could not be denied; they are the outcome of conditions.[18]

In his article in the same issue of the Town Planning Review, Adshead went further: the realities of modern production called for the standardisation not just of components, but of the house itself. With a 'standard cottage', he stated, it would be possible to take advantage of all sorts of 'labour saving contrivances and machine made materials' made possible by mass production. Concrete slabs might be used for walls, which could be strengthened 'by the introduction of a light steel frame to carry the roof' – a clear reference to the Dorlonco system. 'Such a system of construction would be extremely economical if a big repetition could be ensured.'[19]

In pressing the general case for a 'standard cottage to be erected ad infinitum', Adshead parted company with Unwin, who always objected to any standard design – whether of the architect or the speculative builder – and insisted on the role of the architect in preparing individual designs to meet the particular requirements of each scheme and locality[20]. In espousing this course, however, Adshead could look for support from a war-time vogue which, although only loosely related to questions of building construction, nonetheless gave standardisation a considerable attraction in the eyes of the public. This was the movement for industrial standardisation, which was successfully applied in the manufacture of munitions and shipbuilding during the war and which many people were ready to believe held the key to the future of industrial production[21]. By the time of the Armistice 'simplification and standardisation' had become a catch-all slogan. Every problem, it was intimated, was to be solved in the way in which Henry Ford had solved the problems of motor car manufacture. By 1919, MPs were citing the example of the Ford car in relation to the housing programme: 'It must be obvious', they said, 'that standardisation is a good thing'[22].

According to the Liverpool school, however, the standard cottage was a necessity imposed not just by modern production, but also by modern society. Since the turn of the century it had been a commonplace notion that society was entering – not just a new century – but a new phase in its history, 'a new era in political and social life'[23]. This idea that society was entering a new period had been immediately taken up by the nascent town planning profession. Whereas in the nineteenth century, it was claimed, society had been characterised by individualism, people were now realising that the individual must take second place to the good of the community as a whole – as interpreted, of course, by the planner[24]. Unwin advanced this claim in *Town Planning in Practice* (1909) [CD]:

> There is growing up a new sense of the rights and duties of the community as distinct from those of the individual. It is coming to be more and more widely realised that a new order and relationship in society are required to take the place of the old…. The town planning movement and the powers [of planning] conferred by legislation on municipalities are strong evidence of the growth of this spirit of association.[25]

Reilly made a similar point in his article 'The City of the Future' (1910), but introduced a further dimension that was to be developed by Adshead in 1916. The main difference between the immediate past and the period 'on which we are just entering' was, Reilly said, firstly a new standard of taste (of which more later), and secondly the realisation of 'the futility of disorganised individual effort':

> The laissez-faire period of town growth corresponding to the last half of the last century has proved its wastefulness as well as its hideousness; hence our Town Planning Bills and our co-operative suburbs. The note of the new

period, therefore, is organisation – the suppression of rampant individualism for certain general amenities.[26]

Reilly's use of the words 'wastefulness' and 'organisation' betray the influence of what is generally known as the national efficiency movement. The shocks of the Boer War (1898–1902), coming on top of the growing realisation of Britain's economic decline in relation to Germany and the USA, generated among many thinking members of the British establishment ranging from Lord Rosebery to the Webbs what GR Searle has called a 'quest for national efficiency'[27]. Individualism, they believed, had to be replaced by organisation; laissez-faire by the rule of the expert. The idea of 'organising' society on the laws of 'efficiency' was echoed, on an industrial level, in Frederick Winslow Taylor's ideas about the organisation of production. Although concerned only with what went on inside factories, not with society as a whole, Taylor's ideas about efficiency and the way in which it was to be achieved (the importance of science as against tradition and the role of expert analysis in organising the work of others) held much common ground with those of the advocates of national efficiency. Taylor's visit to Britain in 1910 and the publication of his *Principles of Scientific Management* in the following year aroused widespread interest in his views (articles on them appearing in *The Engineer, Engineering, Mechanical World, The Nation*, to name but a few) and when the war came many people were ready to see its 'lessons' in terms of organisation and efficiency[28]. CK Hobson, for instance, wrote in the *Sociological Review* in July 1915 that the war had shown that the 'energy of the nation' must be 'more effectively applied in the processes of production. Rule of thumb methods and slovenly ways of thinking will have to go by the board; scientific training and organisation ... will demand much greater attention than they have received hitherto'[29].

The war seemed to establish beyond doubt the hypotheses about changes in society advanced before 1914. There could be no doubt now that the 'old world of our forefathers'[30] had gone for ever. 'Who does not feel', demanded WH Dawson, 'that since August 1914 England has ... entered an entirely new epoch in her history?'[31] There could be no question now of the importance of national efficiency. Nor, it seemed, could there be any doubt that individualism had at last been replaced by collectivism. With the workings of the free market more or less suspended, and state control replacing individual enterprise as the motor of the economy, even big business was inclined to believe that with the war a new era of collectivism had been born. *The Times* in 1916 called for the permanent replacement of the old 'chaotic world of individualistic business run for unchecked private profit'[32] by a system based on amalgamation, co-ordination and state partnership. Even the new Federation of British Industries – itself a sign of this new attitude – accepted the fact that in modern conditions 'the group, the society and the collective effort are authoritative'[33]. The replacement of individualism by the collectivism hailed by the town planners had, it seemed, finally come about.

It was in this climate of opinion that Adshead advocated the standard cottage as an essential element of the new highly organised form of society. Adopting the quasi-scientific and 'detached' stance of the 'expert', he argued that the standard cottage was a necessity of the new social order:

> The standard cottage is an essential appendage of a highly organised social system, and without it we cannot have that which lies at the very root of national efficiency, organisation and economy.... If the war has proved anything, it has proved that national efficiency in the future will depend almost entirely upon good organisation. Organisation is the keynote of the success not only of the modern nation, but of the modern community. Organisation demands the marshalling of individuals having similar interests, and within limits there should follow uniformity in the appearance of their houses.... [34]

Classical taste

So far, the views of the Liverpool school have mirrored the well-known pronouncements of the modern movement. It is a new age in which, as Budden put it, 'the cry of the prophets of "folk art" has died in the clangor of machinery'[35]. Mass production, standardisation, modern methods of construction, the replacement of the individual client for the mass: Adshead's standard cottage answered the same requirements as Le Corbusier's *maison-type*. How was it then, that the forms they recommended – Neo-Georgian on the one hand, modern on the other – were so different? Followers of the modern movement will certainly ask what intervened to deflect the Liverpool school from what otherwise might have appeared to be the logical conclusion – a new style that would express this new era and these new economic and social realities.

The answer was given by Reilly in the article already noted: the concept of taste. Reilly and Adshead were both disciples of Sir Reginald Blomfield, the leading propagandist of the Edwardian classical revival. As Reilly later recalled, the teaching at Liverpool was 'largely based' on Blomfield ('his books became textbooks for professors and students alike')[36] and it was Blomfield who provided Adshead's testimonial for the Liverpool job in 1909[37]. For Blomfield, architecture meant above all what he termed the 'grand manner' and it was his knowledge of, and training in, the grand manner that defined the architect and gave him his special value. Faced with the realities of the twentieth century, the taste of the architect would lead him to select that period from the Grand Tradition most appropriate to the present. As Blomfield put it in *The Mistress Art* (1908), architecture was 'the art ... which beautifies and clothes building construction': the 'plain facts of construction' had to be transmuted through composition, taste, knowledge of tradition into 'forms that appeal to the imagination by their beauty'[38]. Blomfield himself had a strong preference for the eighteenth century and it was therefore not surprising that it was the 'lesser domestic architecture' of this period – already applauded by Abercrombie and Ramsey – that should be selected by the Liverpool school as the paragon of the new architecture. Having established to his own satisfaction the inevitability of standardised components in a mass housing programme, Lionel Budden pronounced that the only accept-able model came from the eighteenth century:

> A regular recurrence of form is dictated by every valid consideration. What is essential then is that the form shall be good in itself, shall reach a level on which repetition can be sustained. How perfectly this aim was accomplished in the past is demonstrated by the domestic work of the late Eighteenth Century.
>
> The causes which then forced acceptance of the principle of standardisation in Domestic Architecture are a hundredfold more insistent today. They can neither be ignored nor circumvented. Only one right course is open – to see that they lead to beautiful, not ugly, interpretation, to see that they are expressed with distinction, not with vulgarity.... Sound reasons combine to point to late Georgian work as the source from which a large proportion of the motives should be drawn.... [39]

Similarly, when dealing with the standard cottage, Adshead was eager to show that, even when employing his steel-frame system, there would still be oppor-tunity for following the principles of the architecture so much admired by the Liverpool school:

> There will still be scope for embodying in the design of the standard cottage some of that character which we associate with good tradition, and which depends so much upon obtaining characteristic proportions.[40]

The commitment of the Liverpool school to the forms of eighteenth-century architecture pre-dated by several years the formulation of their programme for mass housing. The result was a certain awkwardness in reconciling their analysis of modern society and modern production with their recommendation of an eighteenth-century style of architecture. Here it may not be too fanciful to detect cracks in what was otherwise a solid front presented by the school. Most of them would no doubt have agreed with Adshead's presentation of the case: Neo-Georgian was the appropriate architectural form in which to express the realities of the twentieth century – the correct 'selection' made from the source-book of history.

Budden's argument, however, was slightly different. His disdain for the sordid realities of the twentieth century ('mechanical dreariness ... squalid poverty of mind'[41]) almost led him to suggest that Neo-Georgian was appropriate, not as an expression but as a disguise for contemporary realities – a way of dressing up the unattractive realities of twentieth-century life in the 'quiet and well bred' costume of the eighteenth century. Ever since Pugin, English architects had at least paid lip-service to the belief that architecture should express in an honest and faithful manner the realities that had brought it into being: but if Budden was now recommending the Neo-Georgian as disguise rather than true expression, nobody seems to have objected. Indeed, *The Architects' Journal* thought it was a thoroughly good idea. Reviewing the steel-frame houses at Dormanstown, it noted accurately, and without any hint of disapproval, that 'the finished cottages, which have all the grace and charm of the eighteenth century work, give no indication of the method of construction adopted'[42].

Neo-Georgian versus the modern movement

Today it is probably true that few non-architects would see much to choose between the now stereotyped inter-war 'municipal Neo-Georgian' and the square boxes of the modern movement. Both are generally disliked and, if the former is slightly more acceptable, it is probably only because of its familiarity. Among architects, however, the modern movement product is at least accorded respect because of its basis in theory. By contrast, the Neo-Georgian council house is dismissed as intellectually unmotivated; an ill-considered response to the limita-tions of tight cost controls and government design manuals – certainly not the product of any design theory. It is, of course, true that Neo-Georgian came to be adopted as something very close to the 'official' style for the state-subsidised housing that followed the First World War. By comparison with the elaborate neo-vernacular of many pre-war garden suburbs, the simplified Neo-Georgian offered real cost savings as well as possibilities for the standardisation and mass production of components. These arguments were central to the advice published in the Tudor Walters Report of 1918 [CD] and numerous post-war design manuals; and they were founded in the experience of the largest recent public sector housebuilding programmes, namely, the war-time housing of the Ministry of Munitions (see chapter one). What this study shows is that, on top of the 'common sense and experience' approach of Unwin's circle, there was an ideological basis for the new municipal architecture supplied by the Liverpool school: Neo-Georgian cottages were regarded by the intellectual luminaries of English architecture as the appropriate form for the new age. What we have also seen is that this English theory, whose products were so very different in appear-ance from those of the European modern movement, was astonishingly close to the continental theory in most of its propositions, even if no more articulate in its rhetoric.

Chapter three

An insurance against revolution

It is generally recognised that the Housing Act of 1919 marked a major turning point in the history of public housing in Britain[1]. Whereas in the 25 years before 1914, local authorities accounted for only about two per cent of new dwellings[2], between January 1919 and March 1923 they accounted for more than 60 per cent, and in the inter-war period as a whole for nearly 30 per cent, of new dwellings[3]. Furthermore, the Housing Act of 1919 not only marked a transformation in the scale of public housebuilding; it also marked a similar transformation in its quality. Before 1914 municipal housing had usually taken the form either of high-density tenement blocks on sites cleared of slums, or of cottage estates located in suburban areas where lower land values permitted a more spacious layout. In neither case, however, had there been any particular attempt to exceed the standards of accommodation and design offered by speculative builders.

For the houses built under the 1919 Act, in contrast, the government decreed that a very much higher standard was to be adopted: local authorities were told in the *Housing Manual* of 1919 that 'it is the intention of the Government that the housing schemes ... should mark an advance on the building and development which has ordinarily been regarded as sufficient in the past'[4]. This advance involved both standards and design. The new houses were to conform to the generous space standards laid down by the Tudor Walters Committee [CD], and were to include features such as upstairs bathrooms and linen-cupboards that previously had been found only in the houses of the middle classes. In terms of design, they were to follow not the example of the speculative builder – houses built at maximum densities in long terraces, surrounded by nothing but tar-macadam and brick (Fig. 23) – but the very different model offered by the garden city movement (Fig. 24). The *Housing Manual* specified that estates were to be laid out at low densities – not more than 12 houses to the acre [0.4 hectare] – with gardens, trees and open spaces (Fig. 25), in the manner established by Raymond Unwin at Hampstead Garden Suburb and elsewhere before the war. With design under the control of people such as Unwin himself (chief architect at the Ministry of Health from 1919), the housing schemes built under the 1919 Act were strikingly different from the housing to which the majority of the working class was accustomed (Figs 26 and 27)[5].

At least in outline, all this is fairly well known. It is also well known that the campaign that made these remarkable innovations in both the provision and the design of public housing did not, itself, last for very long. Less than two years

Fig. 23. Typical pre-war speculative builders' housing in Tottenham, north London, c1900

Fig. 24. Parker & Unwin, Asmuns Place, the first artisans' housing at Hampstead Garden Suburb, 1907

Fig. 25. Typical local authority housing built under the 1919 Housing Act: City of York, Tang Hall estate, house plan A103, 1919

after the passage of the Act of 1919, with the number of houses built falling far short of the original target of 500,000, the government brought the campaign to a close, and announced that no further houses beyond those already in tenders approved by the Ministry (approximately 176,000) were to be built under the 1919 Act[6]. At the same time, the Ministry pressed for a reversion to a much lower standard for those houses which were in approved tenders but which had not been started: densities were increased, space standards were reduced, and the luxuries introduced in 1919 were eliminated[7]. This meant that, by 1922–3, such houses as were being built by local authorities were of a type not dissimilar to those erected before the war.

Thus, in terms of both the provision and the design of public housing, the period from 1918 to 1921 is of exceptional interest and importance. How is one to explain a sequence of events in which an extreme lurch in one direction (unlimited Treasury grant, dramatic improvement in design) was followed two years later, with the axing of the housing campaign and the cut in standards, by an almost equally extreme lurch in the opposite direction? For the changes in housing provision, a number of explanations have been advanced. Marian Bowley saw the housing campaign as a direct response by the government to the economic problems for housing created by the war, and suggested that the 'experiment' was brought to an end when it was realised how expensive it had become[8]. Others have suggested that the housing campaign was primarily the work of idealists, and that the inadequacy or sheer impracticality of their vision led necessarily to its demise[9]. Against both of these, Bentley Gilbert argued that the introduction and the demise of the campaign had to be seen in political terms[10]. But none of these explanations has referred at all to the physical form of the houses that resulted[11]. On the contrary the housing campaign has been treated as if it involved nothing more concrete or tangible than other forms of social welfare. The result has been, not only that the changes in design and standards have been left altogether unexplained, but also that the precise nature of the housing campaign has not been identified. For, as will be shown below, the main purpose of the campaign was to prove to the returning 'heroes' and others that there was no need to resort to revolution in order to improve the conditions of life; and the houses were intended to embody this message. The new houses built by the state – each with its own garden, surrounded by trees and hedges, and equipped internally with the amenities of a middle-class home – would constitute visible proof of the irrelevance of revolution.

Preparations for post-war housing

Long before the Armistice of November 1918, government departments in Whitehall were aware of the serious housing problem that would face the country at the end of the war. With the essential resources of capital, labour and materials taken up by the war effort, general residential building (which had produced on average about 75,000 houses per annum before the war) had,

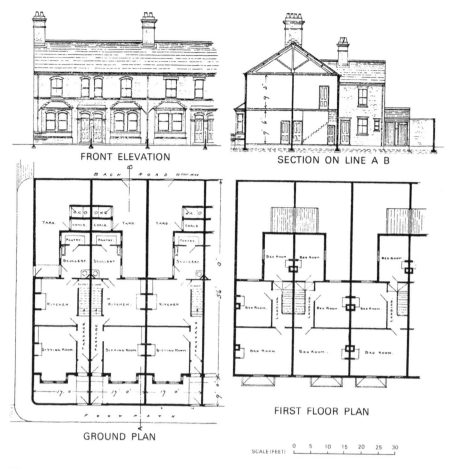

FRONT ELEVATION   SECTION ON LINE A B

GROUND PLAN   FIRST FLOOR PLAN

SCALE (FEET)

Fig. 26. Plans, elevation and section of typical pre-war speculative builders' housing in York, c1900

by the end of 1916, virtually ground to a halt. The result was that, by the time of the Armistice, most parts of the country faced a severe housing shortage. Estimates of this were made for the country as a whole, but its intensity can perhaps best be grasped from the figures for the proportion of unoccupied dwellings on the estates of the London County Council (LCC): this fell from the already low figure of 1.25 per cent in 1914 to less than 0.1 per cent in the early part of 1919[12]. The problem was compounded by the fact that the cost of housebuilding had risen enormously during the war (by at least 100 per cent),[13] and there was a general expectation that it would eventually fall when normal conditions had been restored. A government report stated in 1917:

> In the years immediately following the war, prices must be expected to remain at a higher level than that to which they will eventually fall when normal conditions are restored…. Anyone building in the first years after the war will consequently be faced with a reasonable certainty of a loss in the capital value of their property within a few years[14].

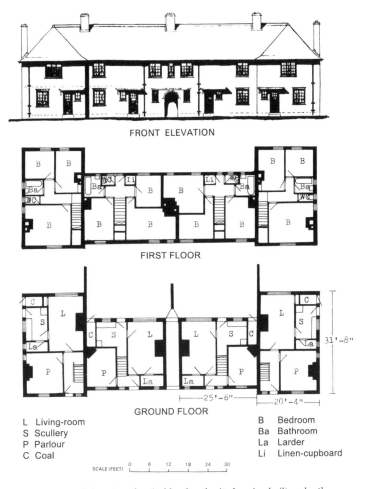

FRONT ELEVATION

FIRST FLOOR

GROUND FLOOR

L Living-room	B Bedroom
S Scullery	Ba Bathroom
P Parlour	La Larder
C Coal	Li Linen-cupboard

SCALE (FEET)

*Fig. 27. Plans and elevation of typical local authority housing built under the
1919 Housing Act: City of York, Tang Hall estate, house plan A104, 1919*

In these circumstances, it was not to be expected that residential building would revive unless the government intervened in some way to offset this loss.

While there was a surprising unanimity on this analysis of the problem (ranging from labour organisations on the one hand to housebuilders on the other), there was no such consensus on the form of its solution[15]. In fact, markedly different attitudes were taken by the two government departments involved. The more conservative view was taken by the Local Government Board (LGB), which, as the body responsible for housing administration before the war, saw post-war provision largely in terms of pre-war traditions: housing would remain the business of local authorities rather than of central government, although the latter might assist the municipalities with some strictly limited form of financial aid[16]. In contrast, the Ministry of Reconstruction was responsible for dealing with the unprecedented problems that would arise with the transition from war to peace, with 'millions of demobilised soldiers, munition workers and other war workers ... suddenly ... at a loose end on the declaration of peace'[17]. If housing was to play a significant part in absorbing this demobilised labour, a very much more dynamic policy would be required: the government would have to require, not simply permit, local authorities to build houses, and in order to get them to do so would have to guarantee them against financial liability[18].

There was a similar disagreement between the two departments over design and standards of accommodation. The LGB considered that the houses built after the war by local authorities should be of the same kind as those built before the war. To make this clear, at the beginning of 1918 the board re-issued, with only the most minor textual alterations, the *Memorandum on the Provision and Arrangement of Houses for the Working Classes* originally published in 1913 (Fig. 28); and it was this conception of post-war housing design that was written into the much-publicised cottage competition held at the same time[19]. But the radicals at the Ministry of Reconstruction were as much committed to a new start in design and standards as in provision; as Seebohm Rowntree commented to the minister, Christopher Addison, 'I believe that the Ministry of Reconstruction takes a very much larger view of the advantage which may be taken of the present situation, materially and permanently to raise the standard of houses ... than is taken by the LGB'[20].

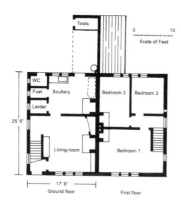

Fig. 28. The largest type plan in the Local Government Board's Memorandum on the Provision and Arrangement of Houses for the Working Classes, *1913*

This difference came out into the open during 1918 when the women's housing sub-committee appointed by Addison produced a detailed and sustained critique of the standards specified by the LGB's memorandum and cottage competition. The LGB responded as follows:

> The Board are not prepared to accept the view ... that it is essential that all houses should have a parlour in addition to a living-room and scullery, nor do they accept the view that a separate bathroom ... is in all cases essential. Moreover they would not be prepared to insist on all houses having three bedrooms....

> Since houses providing three satisfactory bedrooms on the first floor can be designed with a 16 foot frontage ... the Board could not insist on a rigid rule that the frontage should be greater in every case....

> If expense were no object, or if the tenant would be prepared to meet the extra cost ... by the payment of an extra rent, the Board could readily accept the ideal arrangements desired, but as matters stand they are of opinion that less expensive arrangements must be accepted as sufficient[21].

As may be imagined, the report of the women's housing sub-committee did nothing to improve relations between the two departments. Eventually Addison decided that such a 'direct attack upon an LGB publication' could not be published, and the main part of the report was suppressed in the published version [CD][22].

What interested the cabinet, however, was not the details of future housing, or even the housing problem in itself, but rather the role that housing could play in wider political issues, particularly the relationship of government and labour. This can be seen from the events that led to the first announcement of post-war housing policy, in July 1917. In the summer of that year the government faced what appeared to be a major crisis with the 'May Strikes' of the engineering workers and general discontent and war-weariness among the working class, and in June 1917 a commission of enquiry was set up to report on the causes of industrial unrest[23]. The conclusion of the commission was that, among a number of causes of unrest (of which the most important was food), one that was 'acute in certain districts' was the 'want of sufficient housing accommodation'; and it recommended that, even if the government was not prepared to allocate the resources to deal with this by housebuilding, 'announcements should be made of policy as regards housing'[24]. Recognising that 'for the vigorous prosecution of the war a contented working class was indispensable'[25], the cabinet responded immediately to the recommendations made by the commission, and at a meeting on 24 July 1917 (exactly a week after a summary of the commission's findings had been received) authorised such an announcement to be made[26]. Four days later a circular from the president of the LGB informed local authorities of the government's intention of providing 'substantial financial assistance' for housebuilding after the war[27]. As the *Municipal Journal* complained, the terms of the circular were so vague that it was of very little value in terms of the actual preparation of housing schemes; but this was of relatively small consequence to a cabinet concerned less with the practicalities of housebuilding than with the effect of such announcements on public opinion[28].

Armistice and after

Until the beginning of November 1918, the pledges made by the government on both the provision and design of post-war housing conformed to the views of the LGB, rather than the Ministry of Reconstruction. With the Armistice of 11 November, however, the position was transformed. Suddenly the government found itself faced with the emergency foreseen by the Ministry of Reconstruction: the demobilisation of some five million men from the services and the release of a similar number of men and women from munitions' production and other war industries. In this situation the idea of a dynamic housebuilding campaign as a counter to unemployment assumed a new cogency. But the problem of moving from war to peace was not merely one of employment and logistics. To members of the government it appeared that the situation they faced in the aftermath of the Armistice, with 'general unrest and ... the presence of strikers [and] demobilised soldiers in the streets' resembled in all too alarming a manner that which in other countries – Russia and Germany – had led to the overthrow of the state[29].

For, whether or not in the light of hindsight we now consider such fears to have been exaggerated, there is no doubt that, at the time, the cabinet took the threat of revolution very seriously. At the end of January 1919, as the cabinet faced the Clydesiders' strike for the 40-hour week, the Conservative leader and deputy prime minister, Bonar Law, told the prime minister: 'everything depends on beating the strike in the Glasgow area, as if the strikers are successful there the disorder will spread all over the country'[30]. At the same time the secretary of state for Scotland reported to the cabinet that 'in his opinion, it was more clear than ever that it was a misnomer to call the situation in Glasgow a strike – it was a Bolshevist rising'[31]; and Walter Long, the elder statesman of the Conservative party, noted that 'there was no doubt that we were up against a Bolshevist movement in London, Glasgow and elsewhere'[32]. In February the cabinet faced not just the Clydeside strike but the prospect of national paralysis in the key sectors of transport and power, with the underground railway drivers on strike and the electricity workers and miners threatening to follow. The prime minister, Lloyd George, wrote to Bonar Law that, if the miners' strike took place, it would be 'different in character from any hitherto ... a menace to the whole foundation of democratic government'[33]. But while the state appeared thus threatened, the forces on which it relied for its defence seemed less than fully dependable: early in 1919 the cabinet anticipated a renewal of the strike in the police force that had occurred the previous summer, and as regards the armed forces, several mutinies actually broke out in garrisons on both sides of the Channel. In immediate response to the latter, the cabinet decided on 28 January to award pay increases to the armed forces, 'frankly to allay unrest'[34]. Moreover, demobilisation (the real answer to unrest in the forces) presented perhaps the greatest danger of all, for it appeared to the government that demobilised soldiers might form the military vanguard of a revolutionary force. Throughout the first half of 1919 the government was in receipt of alarming indications of discontent and radical tendencies among ex-servicemen's organisations, culminating in July 1919 with the ex-servicemen's boycott of the peace celebrations. It seemed that the mass army that had defeated the Kaiser, now that it was demobilised, might overthrow the government: as the Home Office warned, 'in the event of rioting, for the first time in history, the rioters will be better trained than the troops'[35].

How was this danger to be dealt with? Force, the last resort of government in the past, was clearly to no avail: there was no way that the government could force a demobilised mass army to do anything. The only answer was persuasion. As Lloyd George told Bonar Law in March 1919, it was ideas, not guns, which would decide the outcome: 'the party that secures on its side either general opinion or the opinion of the working classes of the kingdom must win'[36]. On 3 March Lloyd George warned the cabinet:

> In a short time we might have three-quarters of Europe converted to Bolshevism.... He believed that Great Britain would hold out, but only if the people were given a sense of confidence.... We had promised them reforms time and again, but little had been done. We must give them the conviction this time that we meant it, and we must give them that conviction quickly.... Even if it cost a hundred million pounds, what was that compared to the stability of the State?

> So long as we could persuade the people that we were prepared to help them and to meet them in their aspirations, he believed that the sane and steady leaders among the workers would have an easy victory over the Bolsheviks among them[37].

The occasion for this declaration was the consideration by the cabinet of the government's housing bill. In the months following the Armistice the government had promised a wide-ranging programme of social reform (including unemployment protection, hours of work, industrial democracy and land settlement), but at its heart was the promise of a great housing campaign: 'habitations fit for the heroes who have won the war' were promised by Lloyd George on the day after the Armistice[38], a pledge that was regarded – as *The Times* put it – as 'far and away the most important ... in the Government's programme of social reconstruction'[39]. In the terms of Lloyd George's statement to the cabinet, the housing campaign would give the people a 'sense of confidence' in the status quo and prove that there was no need to resort to revolution in order to improve their lot. Compared to the enormous ends that the housing campaign would secure, cost was irrelevant: as the parliamentary secretary put it in April 1919, 'the money we are going to spend on housing is an insurance against Bolshevism and revolution'[40].

Such was the view, not just of Lloyd George and other erstwhile radicals, but of the entire political establishment. In cabinet, the chancellor of the exchequer, Austen Chamberlain, thought they all agreed with the general attitude expressed by the Prime Minister…. He regarded housing as the first problem to be faced…. We ought to push on with it immediately, at whatever cost to the State[41].

Such criticisms of the housing bill as were made by the cabinet were, not that it went too far in getting houses built at public expense, but that it did not go far enough: thus, for instance, the period allowed to local authorities for the preparation of their schemes was reduced from six to three months[42]. Likewise, in the House of Commons a universally favourable reception was given to the government's proposals for housing – proposals that, a year earlier, would have been regarded as unbelievably radical. In the debate on the housing bill (7 and 8 April 1919), one MP after another testified to the belief that the housing programme offered the best hope of a solution to the social crisis. The leader of the opposition, Sir Donald Maclean, spoke for the House when he said:

Unless this vital problem is dealt with promptly and effectively, the social conditions of this country will go from the very serious condition in which they are now to one of which we should shudder to think…. We are face to face with the greatest difficulties…. One of the great difficulties of the future will be unrest, and one of the best ways of mitigating it is to let people see that we are in earnest on this question[43].

If the housing campaign was to fulfil these momentous social objectives, it was clear both that the houses had to be built rapidly, and that they had to be very much better than those of the past. If housing was to be the solvent of social unrest, it could not be left to local authorities to act as and when they pleased: local authorities would have to be required to act immediately, and new powers and new administrative machinery would be needed to ensure that they did. Equally, to induce local authorities to undertake housing schemes in such unsettled conditions as those prevailing (with supplies of building materials and labour uncertain, and with costs already far beyond the amount that could be recouped from the tenant), any elements of financial risk would have to be eliminated, and the Treasury would have to bear responsibility for any loss. It was these points, making such a break with pre-war tradition, that were embodied in the 1919 Housing Act[44].

Similarly, on housing design, ideas that before the Armistice had been regarded as dangerously radical now appeared as nothing other than good sense. If, to use Lloyd George's phrase, the houses were to meet the aspirations of the people, it was no good building houses of the same inadequate standard as those constructed before the war. On the contrary, the houses had to be, and had to be seen to be, as one MP put it, 'on quite different lines' from those of the past, a 'great improvement on anything we have'[45], a point made by no fewer than 10 of the 24 MPs who contributed to the second reading debate on the bill. It was in accordance with this perception that the *Housing*

Manual of April 1919 followed the recommendations of the Tudor Walters Report and required local authorities to break completely with pre-war traditions of design and housing standards[46] (see Fig. 25): as a Ministry official said, 'whatever else these houses may be, they are at least different from pre-war "working-class" houses'[47].

From demobilisation to the demise of the housing programme

The City and the Treasury were opposed to the housing programme from the start, for they wanted the cabinet instead to reduce public expenditure and establish a surplus for loan repayment, in preparation for a return to the gold standard[48]. It soon emerged that, largely due to the government's refusal to operate building controls and to provide capital for housebuilding, the housing campaign was running into serious difficulties, in terms both of the extremely slow rate of completions and the rapid escalation of building costs[49]; and these facts provided ready ammunition for the opponents of the campaign. Throughout 1919 and until the middle of 1920, however, the cabinet ignored or rejected Treasury demands for a cut-back in the housing programme, on the grounds that – as Lloyd George had said in March 1919 – the danger facing the state overruled all questions of financial policy. Thus in May 1920 a Treasury memorandum asked whether a halt should not be called to the housing programme, and Addison, the minister of health, was advised by the first secretary to seek a renewal of support from the cabinet:

> The final justification for the housing programme originally was, I suppose, that it was an insurance against something a great deal worse. It may be that the insurance is still necessary, that the Cabinet are ready to accept the heavy increase in premium.... [50]

The cabinet's decision at this date (June 1920) was that 'in view of pledges given' the proposal to cut back the programme was 'not feasible at present'[51].

In the winter of 1920–1, however, the position was transformed. By the autumn of 1920 it was evident that the post-war economic boom was turning rapidly to slump, with prices falling and unemployment suddenly on the increase[52]. In October a massive demonstration of unemployed workers took place in London, to which the government responded with new measures for unemployment insurance and relief works[53]. It was now relief works for the unemployed, not the threat from labour, that filled the headlines of the national and local press, and by December 1920, as the cabinet noted, it was evident that 'the trade unions were no longer in so strong a position as they had been'[54]. By this date, too, it was apparent that the slump had weaned the ex-servicemen from any ideas of radical politics, and that they were going to establish, not the 'red guard' that had once been feared, but the British Legion[55].

While the growth of unemployment was undermining the strength of labour in general, the cabinet recognised that the 'crucial contest with labour' was that with the miners[56]. By the summer of 1921 this was a battle that the cabinet had joined and won. The autumn of 1920 saw a preliminary skirmish, in which the miners sought the support of the railwaymen and transport workers; but the 'Triple Alliance' – the ultimate threat wielded by organised labour – failed to operate and the miners went out on strike on their own. This was settled indecisively on 3 November, but the conclusion drawn by the government was that the miners had suffered a tacit defeat. Three months later, in February 1921, the cabinet precipitated a final show-down by announcing that the date for the return of the mines to the owners (involving for the miners not just the defeat of their campaign for nationalisation, but also a wage reduction of 40–50 per cent) would be brought forward to 31 March. On that day, accordingly, the miners struck again, and on 8 April the government responded by declaring a state of emergency and mobilising the armed forces. For the government the real danger was not the miners' strike itself, but the threat of a general strike if the Triple Alliance operated; but on the notorious Black Friday (15 April 1921) it failed to do so. Although the miners' strike lasted until the end of June, the crisis for the government was over, and on 18 April the cabinet met to congratulate itself on its successful handling of what already was called the 'recent' industrial crisis[57].

Members of the government were fully aware of the implications of this change in the balance of power for the programme of reconstruction and for the housing campaign at its heart. Since the Armistice, the pattern had been that, while the Treasury called for economy and the reduction of public expenditure, the cabinet was led by what it saw as political imperatives towards generous expenditure on reconstruction, as an insurance against something 'a great deal worse'. Now, however, the danger was receding and the insurance was becoming superfluous. On 29 November 1920 the cabinet finance committee agreed to a Treasury proposal for general cuts in public expenditure, including the imposition of a 'definite limit ... [of], say, 100,000 houses', on the local authority housing scheme[58]. Although he disputed the exact figure, Addison agreed in principle to the curtailment of the programme: for, as he put it, 'it was to be noted that the housing problem assumes a different complexion in present conditions of trade and industry'[59]. Accordingly, at the end of February 1921 Addison instructed his officials that no further tenders for houses were to be approved except where 'a very substantial reduction on past prices is secured'[60]. At the same time he negotiated with the Treasury over the extent of the curtailment; by the beginning of March it was agreed that a present limit of 180,000 houses (the number believed to be in approved tenders) would be imposed, but that when prices had fallen local authorities would be permitted to build a further 70,000[61].

Although it meant a reduction of 50 per cent compared with the original target of the housing campaign, this was not enough to satisfy certain sections of the political establishment. In the changed economic and political conditions, the social reforms so readily promised in the wake of the Armistice took on the appearance of unnecessary and unjustifiable extravagance, for which it was now expected that heads would roll. The Beaverbrook and Northcliffe press launched an 'anti-waste' campaign, attacking 'squandermania' in general and Addison and the housing programme in particular[62]. At the end of March 1921 Lloyd George attempted to deflect the attack by moving Addison from the Ministry of Health, but in June, following another by-election defeat of a government candidate by an anti-waste campaigner, he decided that more dramatic measures were needed. On 10 June Lloyd George wrote to Austen Chamberlain that the by-election result at the St. George's division of Westminster was 'a great warning: 'the middle classes mean to insist upon a drastic cut-down: nothing will satisfy them next year except an actual reduction in taxes'[63]. Two years earlier, political considerations had led the cabinet to lavish expenditure on housing, but now with squandermania rather than the threat of revolution as the dominant consideration, they pointed in the opposite direction. At the end of June unemployment stood at over two million and, as the cabinet noted (29 June), 'now that the coal dispute was settled ... [there was] not much danger of active unrest in Great Britain'[64]. Responding to Treasury proposals for a further round of expenditure cuts, the cabinet finance committee decided on 30 June to bring the housing campaign to a final and conspicuous halt: 'in view of the difficult financial situation', it was decided, 'there was no alternative open to the Government but to decide housing questions not on merits, but on financial considerations only'[65]. Two weeks later, on 14 July 1921, the new Minister of Health, Sir Alfred Mond, announced in the House of Commons that expenditure on local authority schemes under the Housing Act of 1919 would be limited to the 176,000 houses already in approved tenders[66]. Allowing for the time-lag involved in building

Fig. 29. *Typical Ministry of Health design from* Type Plans and Elevations, *May 1920*

operations (which meant that many of the houses for which tenders had been approved would not be completed until 1922 or 1923) this meant the end of the 'homes fit for heroes' campaign.

At the same time and for the same reason, housing standards for houses in approved tenders, which were yet to be started, were sharply cut. Until the end of 1920, despite the enormous escalation in costs, the Ministry had maintained the standards of accommodation and design specified by the Tudor Walters Report (Fig. 29)[67]. But once the decision had been taken to curtail the programme, standards fell sharply: from the beginning of 1921 the Ministry pressed local authorities to adopt smaller and cheaper types of houses, which economised both on the size and number of rooms and on the level of equipment (Fig. 30)[68]. These houses were no longer intended to prove to the people that their aspirations would be met under the status quo: indeed, in so far as they contained any directly political message, it was something rather different. As the *Municipal Journal* noted approvingly in July 1922, these small houses (Fig. 31) could be taken as proof that government and local authorities were no longer 'led astray by visionaries', but were simply getting on with the job of supplying 'the

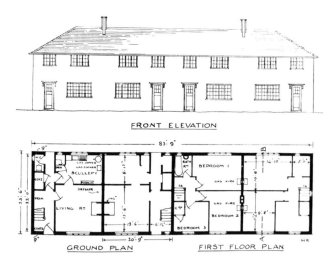

FRONT ELEVATION

GROUND PLAN   FIRST FLOOR PLAN

Fig. 30. Economical design 1921: design published by the Ministry of Health, January 1921

Ground floor   First floor

Scale of Feet

Fig. 31. London County Council, typical narrow-fronted plan from the set of type plans adopted in February 1922, based on pre-war designs

cheapest form of housing which will actually provide accommodation for the poor'. These small two-bedroom houses, it was now claimed, were 'more likely to solve the housing question than model houses with parlours and palace-like amenities in picturesque surroundings.'[69]

Housing and the battle of opinion

The changes that took place in the provision and design of public housing between 1918 and 1921 derived primarily from political considerations. In 1919, faced with what seemed a political crisis of the most serious kind, the state put its faith in the housing programme to win the loyalty of the 'heroes' and the people. With ex-servicemen and others all over the country unable to obtain housing accommodation, it appeared to the government that decisive action on housing offered the best hope of winning the 'battle of opinion' on which everything depended. As the cabinet was told after the partial boycott of the peace celebrations by disaffected ex-servicemen in July 1919, the belief was that 'discontent would be greatly allayed if ... there were some ocular evidence that housebuilding was in progress'[70]. The houses built by the state on garden city lines, with their gardens, bathrooms and other amenities, were intended to prove that the luxuries previously confined to the middle classes were now to be enjoyed by all. In every urban and rural district of the kingdom, the new houses – unmistakably different from the working-class houses of old – would provide visible evidence that a great improvement in the conditions of life could be achieved under the existing system, without any need for revolution.

Had there been a surplus of houses available at the time, the promise to build a further 500,000 (albeit of a greatly superior kind) could scarcely have exerted the influence for which the government hoped: it was the seriousness of the material problems of housing that gave the promise of the housing campaign its ideological potency. But this does not mean, as Bowley and others have implied, that the housing campaign was nothing more than an innocent response by the government to the housing problems bequeathed by the war[71]. Such a view is incompatible both with the evidence presented by the design of the houses themselves, which indicates that there was a much more deliberate intention at work, and with the evidence of cabinet and other government papers. It was not the housing problem per se that, at a time when the City was calling for massive spending cuts, led the cabinet and the House of Commons to allocate unknown but undoubtedly enormous sums for housebuilding. Nor was the cabinet motivated by any feeling of idealism – a desire to put the world to rights after the horrors of war – as has also been suggested[72]. On the contrary, the reason that, in March 1919, the chancellor of the exchequer wanted the housing campaign pursued 'at whatever cost to the State' was the same as that which made the cabinet ignore the call for expenditure cuts in general – the pressing danger of social upheaval and the overwhelming need for the state to validate its claim to the loyalty of its citizens[73].

Similarly, it was not so much idealism as a sure grasp of political reality that led the state to adopt new standards of design and accommodation. Until November 1918 the LGB was firmly opposed to any major improvement in housing standards, because of the cost involved. But given the ideological role assigned to the housing programme in the wake of the Armistice, there could be no question of building the old type of houses when so clearly a superior alternative was available as that offered by the garden city movement and the Tudor Walters Report. If, to use Lloyd George's phrase, the houses were to meet the aspirations of the people, there was only one conclusion for housing standards: as members of parliament recognised, the new houses had to be 'on quite different lines' from those of the past.

In 1919 these innovations in the provision and design of public housing were adopted as an insurance against revolution. Two years later the danger had disappeared and both were abandoned. It was not that the housing problem had been alleviated: in quantitative terms the housing shortage was, if anything, worse in 1921 than it had been two years earlier, and, although costs had fallen, they were still far from the level at which working-class housebuilding could become profitable for private enterprise[74]. As has been seen, the cabinet was fully aware of this and did not even attempt to justify its decision to axe the housing programme on these grounds: the reason for the decision, it stated, was purely fiscal[75]. This emphasis on cost and finance has led some historians into thinking that it was the increase in the cost of the housing programme that brought about its demise[76]. This was not the case. In fact, while building costs rose enormously between 1918 and 1920 (by an estimated one hundred per cent), they actually fell, by about fourteen per cent, in the nine months prior to July 1921[77]. What really distinguished the situation in 1921 from that which had prevailed during 1919 and the first half of 1920 was not the cost of housing, but the unwillingness of the state to bear that cost. In 1919, as Bolshevism swept through Europe, the cost of the housing programme had been of no account; but by 1921, the danger against which housing had offered an insurance had receded, and the cost of the premium appeared as the insuperable objection to an insurance that was now regarded as superfluous.

It was this changed outlook that was also responsible for the fall in housing standards after 1921. On a priori grounds, it might have been expected that, in the period 1918–1920, the Ministry would have abandoned the high standards recommended by the Tudor Walters Report as an obvious way of countering the soaring increase in building costs. This, however, did not happen: in all major respects – room sizes, number of rooms, levels of equipment, housing density – the Ministry maintained the standards of the Tudor Walters Report. It was only from the end of 1920, as the government perceived the housing programme to be no longer indispensable, that the Ministry started to encourage substantial departures from Tudor Walters standards – a process that accelerated thereafter. In other words, the upstairs bathrooms and other luxuries of 1919 were eliminated only when the insurance that they offered was no longer needed.

What is the significance of this study? At the end of the First World War the British government discovered that, in addition to its obvious material uses (as a necessary part of economic life), housing could be used for a rather different purpose – to influence people's ideas about society. It was this discovery of the ideological potential of housing that was responsible for the major changes in both the scale and quality of public housing that ensued. In carrying out this ideological function, the design of the houses was of crucial importance: as MPs repeatedly said, it was the physical form of the new houses that would prove that revolution was superfluous. The implication of this particular episode in the history of housing is clearly that those who wish to understand the history of housing cannot afford to ignore questions of design, and that on the contrary they must include the physical form of the houses within their field of study. The historiography of 'homes fit for heroes' is instructive at a more general level as well: it suggests that the lack of attention given to design by most historians may be a function less of the true importance of design in the real world, than of peculiarities in the training of the historian.

Chapter four

Rationality and rationalism

It is a curious fact that none of the classic English-language histories of modern architecture acknowledges any direct contribution by English architects to the making of modern architecture. Nineteenth-century figures like Philip Webb may be given an important place as 'precursors' or 'forerunners', but once we reach the key period after 1900 the English names disappear. The picture was established by Pevsner in *Pioneers of the Modern Movement* (1936): English architects were given a leading role in the nineteenth century, but after 1900 'England dropped out and Germany took the lead'[1]. The same story appeared in *An Outline of European Architecture* (1943): 'for the next forty years, the first forty of our century, no English name need be mentioned'[2]. While some English-language histories of modern architecture have questioned other aspects of Pevsner's interpretation, on this question of the non-involvement of English architects in the key developments, the interpretation has hardly been modified. Neither Giedion (1941) nor Hitchcock (1958) allowed any direct participation by English architects in the 'new architecture'[3]. Reyner Banham revised the Pevsner interpretation in many respects, but left this one unchanged: English architecture makes an early exit from *Theory and Design in the First Machine Age* (1960), dismissed with the memorable phrase 'a singular example of failure of nerve and collapse of creative energy'[4]. What is omitted by this interpretation is, principally, the social architecture of the garden city movement. Schemes such as Hampstead Garden Suburb (1905–) designed by Raymond Unwin are either excluded altogether (Giedion, Banham) or admitted only as the tail-end of the decline of English architectural progressiveness (Pevsner, Hitchcock). They are most definitely not admitted, as they were seen by those involved at the time, as direct and vital contributions to the rethinking of architecture that took place in the first 30 years of this century. This position has not been altered by subsequent English-language histories: in Jencks (1973), for instance, there was no mention of Unwin or Hampstead Garden Suburb, while Frampton (1980) mentioned Hampstead Garden Suburb, but only as an example of 'enfeebled' design[5].

It has taken an Italian critic to produce a general history of modern architecture which allows a place for English social architecture. Manfredo Tafuri declared in 1974 that modern architecture should be seen less in terms of language than in terms of production. In this view the central lineage was that of architect as social reformer and urban manager, beginning with Unwin before the First World War and encompassing Clarence Stein and Henry Wright in America and Ernst May,

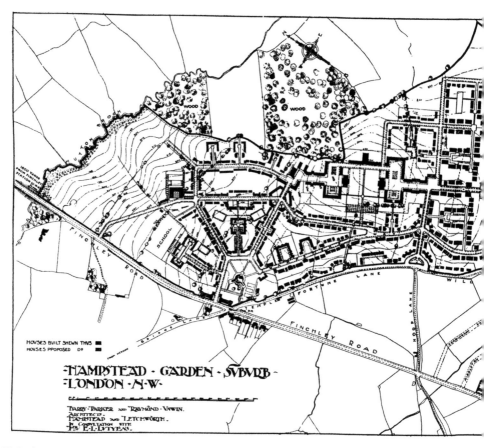

Fig. 32. Parker & Unwin, Hampstead Garden Suburb, site plan, from Town Planning in Practice, 1909

Hannes Meyer and Martin Wagner in Germany in the 1920s[6]. In other words, far from signalling the end of the English contribution to the pre-history of modern architecture, garden city schemes like Hampstead Garden Suburb were to be seen as the starting-point of modern architecture. Tafuri attempted to carry out this reinterpretation in *Modern Architecture* (with F Dal Co, first published in Italian in 1976), but the result was seriously marred both by the a-priority of much of the argument and by historical inaccuracies[7].

This somewhat curious story of the history of modern architecture provides the context of the present enquiry. The subject is the relationship between the theory and practice of the new architecture of the 1920s and that of the pre-war social architecture of the British garden city movement. The latter is exemplified in the designs and writings of Raymond Unwin; the former in the 1929 (Frankfurt) and 1930 (Brussels) meetings of what is called in German the *Kongress für neues Bauen* and in French and English CIAM (*Congrès Internationaux d'Architecture Moderne*). The focus is on an aspect of theory and practice that, quite literally, underlay any question of form: the relationship of building to site.

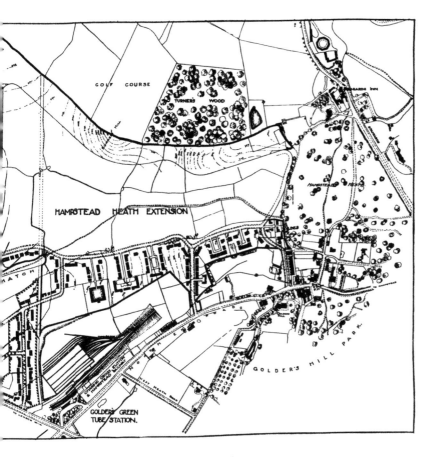

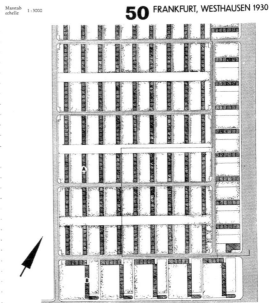

Masstab
echelle   1 : 3000

50 FRANKFURT, WESTHAUSEN 1930

Fig. 33. Ernst May and others,
Westhausen Siedlung, Frankfurt,
site plan, from CIAM, Rationelle
Bebauungsweisen, *1931*

At first glance nothing could be more different than the relationship of building to site at Hampstead Garden Suburb (Fig. 32) and at a typical CIAM scheme such as Frankfurt-Westhausen (Fig. 33). But were they really so different? And was the thinking behind them as unrelated as the difference in configuration would suggest? The answer to these questions involves both the thinking of the new architecture and its relationship to other traditions in architecture: what is at issue is our understanding of modern architecture and the theories on which it is based.

Unwin: site planning in theory and practice

Unwin's career is well known and does not need to be recited here[8]. What follows is an outline of the theory and practice of Unwin's site planning, as evidenced both in his voluminous writings and on the ground[9]. For the latter, Hampstead Garden Suburb, as the most internationally renowned of Unwin's schemes, will be taken as exemplar.

Unwin's overriding concern was to provide a desirable residential environment, not just for the wealthy, but for all. This meant an equal concern with quality and with economy: the problem was, precisely, to provide an environment that was both desirable and within the means of the ordinary family. Desirability for Unwin meant chiefly open space and greenery – in other words, low density; economy was to be achieved by savings in the cost of constructing roads and mains services. The famous diagram from *Nothing gained by Overcrowding!* [CD] showed how savings in road construction could help offset the higher land costs involved in a low-density scheme (Fig. 34). In practice, as at Hampstead Garden Suburb (eight houses per acre [0.4 hectare] overall), road widths were minimised

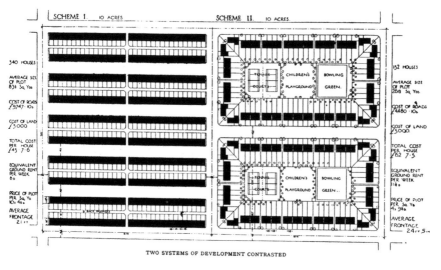

Fig. 34. Raymond Unwin, conventional versus low-density layout, from Nothing Gained by Overcrowding!, *1912*

and full roads provided only where absolutely necessary. The basic technique involved the creation of large blocks of land ('superblocks') with houses ranged along the perimeter and the backland opened up by a number of specific devices. These included the cul-de-sac in various shapes and forms, and greens which were placed adjacent to the road with houses along three sides (Fig. 35). In both of these a track or path would provide access to the houses: in other words, Unwin was saving on road-construction costs by separating pedestrian from vehicular routes, on the basis that it was unnecessary for every house to face directly and fully onto a properly constructed road.

Fig. 35. Parker & Unwin, a quadrangle at Hampstead Garden Suburb, from Town Planning in Practice, *1909*

Unwin defined the job of the architect as the creation of beautiful surroundings. Here he drew extensively from his reading of the Viennese theorist Camillo Sitte, particularly in the notion that beauty was to be found in enclosed spaces (see chapter eight). At Hampstead Garden Suburb and in his major text, *Town Planning in Practice* (1909) [CD], Unwin conceived road planning in terms of the creation of a series of street pictures (see chapter eight). Another major influence on the site plan was orientation or aspect. Unwin shared the universal belief that sunlight was essential to health. 'Let no house be built with a sunless living-room', he wrote in *Cottage Plans and Common Sense* (1902)[10] [CD]: not only was the internal planning of the house to be organised around aspect, but the placing of the buildings on the site had to take this into account as well.

In *Garden Cities of Tomorrow* (1898/1902) Ebenezer Howard had proposed that the desirable environment could be created only in an entirely new city. Unwin, however, believed that a more practicable method was to build not satellite cities but satellite suburbs, attached to, but separated from, an existing city. De facto this was the position at Hampstead Garden Suburb, which was separated from the rest of London by the open space of Hampstead Heath. Unwin formulated this as a general model in his 1911 Warburton Lecture[11]. Under the title 'The Garden City Principle applied to Suburbs', he depicted a city surrounded by a ring of satellite suburbs, each separated from the city and from one another by open countryside (Fig. 36).

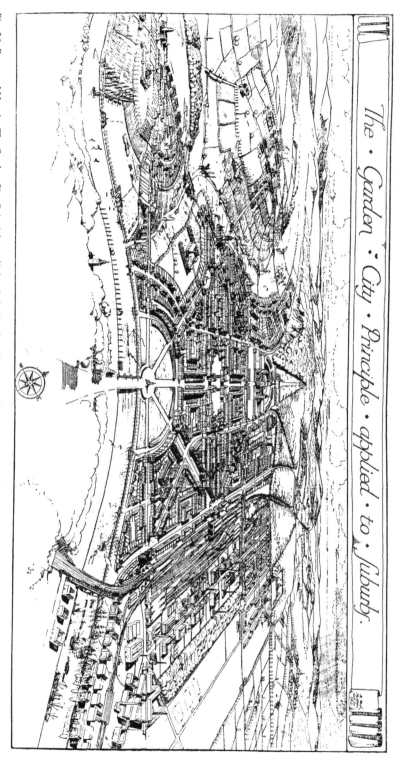

The · Garden · City · Principle · applied · to · suburbs.

Fig. 36. Raymond Unwin, The Garden City Principle applied to Suburbs, from The Town Extension Plan, 1912

To this description must be added one further, and fundamental, character-istic of Unwin's site planning. Unwin insisted that design was not something to be imposed by the *diktat* of the architect, but was to be derived empirically from the given elements of the site, context and needs of the inhabitants. Thus in his articles on the planning of Hampstead Garden Suburb he described how the site plan was arrived at from consideration of data such as the lie of the land, the need for access to the underground station at Golders Green, and the desir-ability of making best possible use of the view over the heath. The survey was of fundamental importance, because the plan was to be derived to a large extent from what the survey revealed[12]. Detailed questions of design were to be treated in a similarly empirical way. Important points in the plan (entrances, bounda-ries, centres) were to be given a treatment that expressed their special status: the centre should have a distinctive feel, and the edge of the suburb should not be left un-noted, but should be marked in some definite way (as by the famous wall along the boundary with the heath). In this way, Unwin believed, design would express the 'realities' – topographical as much as social – from which it was derived (Fig. 37).

Fig. 37. Parker & Unwin, the 'great wall' at Hampstead Garden Suburb, from Town Planning in Practice, *1909*

At the heart of Unwin's work lay his antipathy to the practices of the specu-lative builder. This was itself a part of that wider hostility to capitalists and capitalism that he had imbibed from William Morris and other socialists in the 1880s: 'Our towns and suburbs express by their ugliness the passion for gain which so largely dominates their creation', he wrote[13]. For Unwin, a basic char-acteristic of speculative (in his terms, profit-based) housebuilding was the use of the same inflexible format for both layout and internal planning, irrespective of the particular circumstances or the needs of the inhabitants. In place of this, Unwin saw the job of the architect as producing something that met the needs of the inhabitants and corresponded with the particular conditions (of site, prospect, aspect) prevailing. Good design meant responding to the individual requirements of each location: 'taking advantage of all the opportunities offered by the site, position and aspect of each house in order to secure the greatest comfort and so obtain the best value for building cost'[14].

European absorption of Unwin

By the beginning of the 1920s, what Unwin had been doing at Hampstead Garden Suburb and elsewhere was well known to European architects; indeed, it was often taken as the starting-point in residential site planning. Many European architects had visited Hampstead Garden Suburb (as for instance at the 1910 RIBA Town Planning Conference); others had read the translations of Unwin's writings (*Town Planning in Practice* was translated into German only a year after the English original). Out of the multitude of examples of this European absorption of Unwin that could be cited, two or three will have to suffice.

In 1917 a small housing scheme, St Nicholas d'Aliermont, was designed for a Normandy munitions manufacturer (Figs 38 and 39). The architect was C-E Jeanneret, yet to make his name as Le Corbusier. Before the war he had been studying residential planning and had drafted a manuscript on *La Construction des Villes* which drew heavily for both form and content on Camillo Sitte. The chapter on city design cited Hampstead Garden Suburb as the ideal[15]. The 1917 housing project was what might be expected from this background: on a small, narrow site Jeanneret/Le Corbusier deployed a T-shaped cul-de-sac, with houses arranged picturesquely around it and a second exit arranged off-axis. It was essentially a Sittesque reading of Unwin[16].

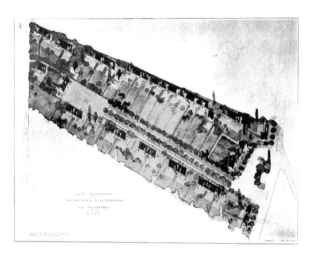

Fig. 38 C-E Jeanneret (Le Corbusier), St Nicholas d'Aliermont, Normandy, 1917, (left) bird's eye view, and Fig. 39 (below), elevations

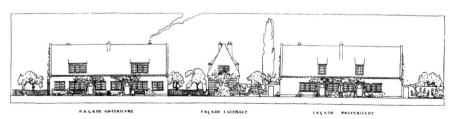

French absorption of Unwin was not confined to Le Corbusier. In 1919 a plan for a ring of satellites around Paris was drawn up by Henri Sellier of the *Office public de la Seine*; in his report on design (1919) Sellier quoted freely from Unwin and reproduced key diagrams from *Town Planning in Practice* and the Tudor Walters Report [CD] (see chapter five)[17]. The site planning of the resultant *cités-jardins* (such as Suresnes, designed in 1920) was based on Unwinite practice[18]. In Germany the picture was similar. For instance, Paul Wolf was a member of the planning faculty of the Technische Hochschule in Charlottenburg, Berlin, who advocated both the satellite-suburb idea (*trabantenprinzip*) and a manner of detailed design of both layout and buildings that drew heavily on Unwin[19]. Another proponent of Unwin's ideas was Ernst May, who had worked with Unwin at Hampstead Garden Suburb before the war. May's competition entry for the city plan of Breslau (1920) was a clear demonstration of the satellite-suburb idea; and when he was appointed *stadtbaurat* of Frankfurt in 1925, May produced a city plan based on the satellite principle, with an elongated suburb to the north-west of the city, detached from it by the open space of the Nidda Valley (Fig. 40). May signalled his debt to Unwin in the first issue of *Das neue Frankfurt* (1926), which carried an article by Unwin on 'the new city'[20].

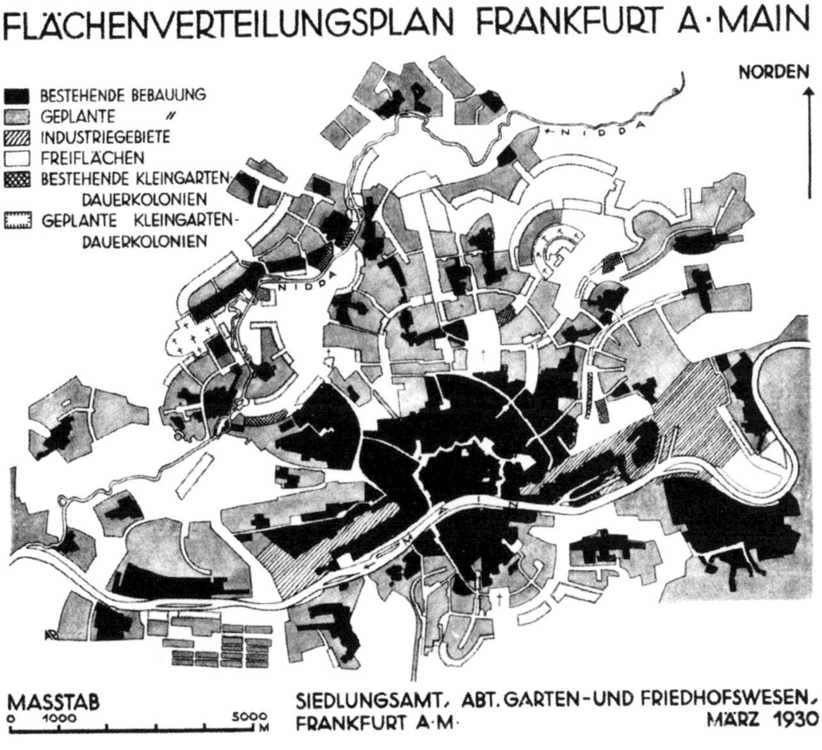

Fig. 40. Ernst May and others, Frankfurt development plan, 1930

The extent to which May was prepared to follow Unwinite practice in site planning can be seen at one of the most famous schemes in the Nidda Valley project: the Römerstadt Siedlung, designed in 1927 (Fig. 41). The site was a narrow strip between the main road to the north-west and the river valley to the south-east. The site plan ran the residential roads along the contours, parallel to the river, and bisected them with a transverse road containing shops and communal facilities. This central road was formed of two successive curves, in the manner of the picturesque followers of Sitte, and in the residential roads the street picture was carefully controlled in an equally Sittesque manner: on the eastern side by curves; on the western side by displacements in the building line. The view into the residential roads was closed by sharp deflections in the course of the road. In both halves of the scheme the buildings on the river side were interrupted by footpaths giving access to belvederes, which were themselves tied in to the rampart that ran along the edge of the suburb, overlooking the river. All of this was consistent with the principles that May would have learnt from Unwin at Hampstead Garden Suburb: the plan was derived from site; the street picture was tightly controlled; and due deference was paid to the prospect over the river valley. The rampart marked the edge between suburb and countryside in the same way as Unwin's wall at Hampstead Garden Suburb[21].

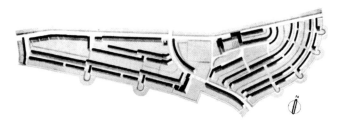

Fig. 41. Ernst May and others, Römerstadt Siedlung, Franfurt, site plan, 1927

But it should be noted of Römerstadt that, while the planning principles were largely in accordance with Unwin, in at least one respect they were at variance: the density was much higher. As German commentators complained, high land prices in Europe made it economically unviable to adopt Unwin's very low densities. At Römerstadt the higher density meant that the superblock had been lost: gardens were small and, with every dwelling fronting onto a fully made-up road, the proportion of the site given over to roads was high – with a direct effect on costs. Analytically, we might say that Römerstadt was back to the 'bad half' of Unwin's Nothing gained by Overcrowding! diagram (Fig. 34): high density meant a lot of road and a paucity of garden space.

CIAM: site planning in theory and practice

At the start of the 1920s Unwinite practice represented the orthodoxy in site planning. By the end of the decade, when the third meeting of CIAM took place

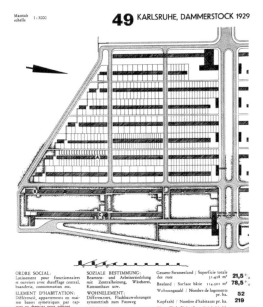

Fig. 42. Walter Gropius, Dammerstock Siedlung, Karlsruhe, site plan, from CIAM, Rationelle Bebauungsweisen, 1931

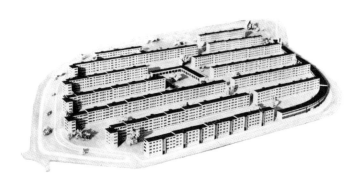

Fig. 43. Otto Haesler, Rothenburg Siedlung, Kassel (southern portion), model, 1930

on the subject Rational Methods of Site Planning (Brussels 1930), the orthodoxy was, to all appearances, very different (Figs 33 and 42)[22]. Instead of subtle adjustments of building to site and careful composition of street pictures, the buildings were placed in long parallel rows with uniform orientation, and the roads ran in equally uniform fashion at right angles to the buildings. This type of layout, known as *Zeilenbau*, formed the basis of the debate over building height that dominated the Brussels meeting, with Gropius arguing for a reduction in the number and increase in the height of the rows, and it also predominated in the site-planning exhibition that accompanied the congress[23]. By this date *Zeilenbau* schemes in Germany included those by Gropius (Karlsruhe-Dammerstock, Berlin-Siemensstadt etc), Haesler (Kassel-Rothenburg etc) (Fig. 43) and May

(Frankfurt-Westhausen). This last was designed in 1929 for a flat site only a short distance down the river from Römerstadt and illustrated in a striking manner the changes that had taken place in site planning.

The whole layout of Westhausen was rigidly geometrical, with both buildings and roads aligned on a rectilinear grid (Fig. 33). The basic technique involved the separation of both buildings from roads and pedestrian from vehicular circulation by placing the rows of dwellings at right angles to the roads. The four-storey blocks of flats along the eastern side of the scheme were aligned east-west; each block was entered from an access path on the north side, with the area to the south of each block reserved for gardens. Similarly, the two-storey terraces that comprised the major part of the scheme had access paths running down the east side of each row, with the west side used for gardens. Thus not only did every dwelling have the same aspect, but each was entered from the same side – which meant that the internal plan and arrangement of the dwellings could be completely standardised. The rows did not run continuously across the superblock, but were interrupted by bands of parkland running east-west across the site. The road system consisted of a grid of distributor and access roads. Overall, the result of this arrangement was that, despite the fairly high density (26 dwellings per acre [0.4 hectare]), only a very small proportion of the site was taken up by roads – a marked contrast to Römerstadt[24].

The *Zeilenbau* system thus offered obvious savings in road construction, and undoubtedly this was seen as one of its main advantages. The social housing programme in Germany was under sustained attack for catering only for the well-off, and every effort therefore had to be made to reduce costs. The device of running the buildings at right angles to the road was well known as an economical format – for instance, to those responsible for the layout of barracks during the war – and it was recommended on these grounds in the Tudor Walters Report of 1918[25]. During the 1920s it was taken up in Germany and was to be seen at schemes designed by Fischer in Munich and Haesler in Celle[26]. Catherine Bauer reported that the figure of 15 per cent was widely quoted as the saving in development costs obtainable with the *Zeilenbau* method[27].

But *Zeilenbau* was not just cheap: it was also seen as the most desirable form of layout. Barbara Miller Lane has pointed out that it was seen as providing the garden city advantages – open space, gardens, parks, trees – at a reduced cost: by 'giving the buildings more land or seeming to do so, [it] introduced a suburban feeling ... and in this respect carried out the aims of the pre-war garden city movement'[28]. An earlier American commentator Catherine Bauer, perhaps with an eye to a Radburn audience[29], pointed to the benefits of *Zeilenbau* in terms of distancing dwellings from the noise and dangers of traffic: the separation that Unwin had given to some of the dwellings (with the cul-de-sac or green), *Zeilenbau* gave to all[30]. But undoubtedly the greatest attraction of the new form of layout was in terms of orientation: the fact that every dwelling could have the best possible insolation.

From the early 1920s systematic studies of insolation patterns had been undertaken in Germany which pointed to the advantages of uniform orientation. At Frankfurt, Walter Schwagenscheidt (appointed by May in 1928) undertook 'scientific' investigations which showed that the best insolation was achieved when the building was aligned at 22.5° from north-south. This was the over-riding attraction of *Zeilenbau*: every dwelling was given this scientifically-proven optimum insolation[31] (here again, to Catherine Bauer's eyes, the German system ensured for all the benefits that Unwin's system could offer to only some). Accordingly, to the Frankfurt architects *Zeilenbau* was, quite simply, both 'the most economic method ... and the one most beneficial to the community'[32]. Gropius summarised its advantages in his address to the CIAM Brussels meeting of 1930, in which (to the chagrin of the Frankfurt architects) he called for high-rise blocks (Figs 44 and 45).

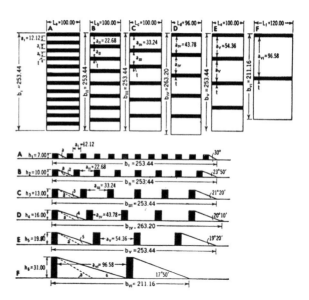

Fig. 44. Walter Gropius, 'Low, Middle or High Rise?', 1930

Fig. 45. Walter Gropius, project for 11-storey slab blocks on a Zeilenbau layout, 1931

Parallel rows of apartment blocks have the great advantage over the old peripheral blocks that all apartments can have equally favourable orientation with
respect to the sun, that the ventilation of blocks is not obstructed by transverse
blocks, and that the stifled corner apartments are eliminated. Such parallel
rows also provide for the systematic separation of highways, residential streets
and footwalks more easily and at less cost.... It makes for better illumination
and more quiet, and also decreases the cost of road building and utilities
without decreasing the effectiveness of land use. The overall distribution is
thus considerably functionalised, resulting in improved conditions of hygiene,
economy and traffic[33].

As the last sentence (with its use of the word 'functionalised') indicated,
however, for the CIAM architects the attraction of *Zeilenbau* consisted of more
than just the sum of these individual benefits. For them, it was part of a far more
ambitious programme, the *Neues Bauen* (New Building), which involved nothing
less than the transformation of building and housing production. The production of dwellings was to be re-cast on the basis of the most modern American
techniques – principally those of FW Taylor and Henry Ford. Taylor's method,
as set out in *The Principles of Scientific Management* (1911), involved the abolition
of traditional or, as he saw them, irrational working practices and their replacement by new, 'rational' procedures based on scientific analysis and calculated
for maximum efficiency. Ford stood both for assembly-line production, which
he applied to automobile production in 1912–13, and for the philosophy of
pre-production research and the 'final type' (in his case the Model-T, which he
regarded as the definitive, scientific solution to the problem of the mass-produced
car). Together, Ford and Taylor were seen as the apostles of rationalisation, a
system which, by increasing efficiency and output, was the answer to all problems at once: reducing the price to the consumer, increasing the profits of the
manufacturer and raising the wages of the worker. As Ford put it in *My Life and
Work* (1922), 'everyone who is connected with us – either as a manager, worker
or purchaser – is the better for our existence'[34]. Ford's book was translated into
German in 1923 and became a best-seller: to a Germany ravaged by war and
economic crisis it seemed that American production methods were as vital to
recovery as the American capital advanced under the Dawes Plan of 1924.

Architects shared this belief in the rationalisation of production (see Fig. 2).
The La Sarraz declaration of the founding meeting of CIAM (1928) stated
that:

The most efficient method of production is that which arises from rationalisation and standardisation. Rationalisation and standardisation act directly on
working methods both in modern architecture (conception) and in the building
industry (realisation).[35]

As Le Corbusier wrote, what was needed was 'a house built on the same principles as the Ford car', or what the Germans called a *'Wohn-Ford'* ('Ford-dwelling')[36]. In an article on the 'Industrial Production of Dwellings' (1924), Bruno Taut stated that 'the problem of housebuilding must be tackled along lines that are valid in industry for the production of machines, cars and similar objects'. Referring to the 'success of Henry Ford', Taut stated that 'exactly the same can be applied to house building'[37]. In 1927 an article by Gropius set out a programme for a 'rational construction industry', in which the old ways of building (unreliable, inaccurate, ad hoc) would be replaced by a new system of 'rationalised management' and the 'mass-production of off-the-peg housing'[38]. In this scheme the architect's job was that of organising production in order to replace all vestiges of the old, irrational practices with the new, scientific system. This programme was taken up by the Reichsforschungs Gesellschaft für Bau und Wohnung (RFG), established in 1927, and put into practice on schemes such as Dessau-Torten and Spandau-Haselhorst. Similar notions informed May's work at Frankfurt, where a 'housing factory' was set up in 1926 on the model of a modern industrial process complete with pre-production trials and prototypes[39]. Gropius summarised the theory, albeit in rather bland form, in *The New Architecture and the Bauhaus* (1934):

> Building, hitherto an essentially manual trade, is already in course of transformation into an organised industry. More and more work that used to be done on the scaffolding is now carried out under factory conditions far away from the site....

> And just as fabricated materials have been evolved which are far superior to natural ones in accuracy and uniformity, so modern practice in house construction is increasingly approximating to the successive stages of a manufacturing process. We are approaching a stage of technical proficiency when it will become possible to rationalise buildings and mass-produce them in factories....

> Ready-made houses ... will ultimately become one of the principal products of industry. Before this is practicable, however, every part of the house – floor-beams, wall-slabs, windows, doors, staircases, and fittings – will have to be normed....

> The outstanding concomitant advantages of rationalised construction are superior economy and an enhanced standard of living[40].

In the eyes of the CIAM architects, *Zeilenbau* was the form of site plan that corresponded to this new era of rational production. Unwin's method of site planning necessarily involved a whole range of house plans, each suited to the vagaries of its orientation, but with *Zeilenbau* the same plan could be used repeatedly and hence mass-produced. Furthermore, the design itself would be optimised in the manner of the Ford car, for the repeated production of a standard design both necessitated, and provided the funds for, research to ensure optimal quality. 'Just as industry submits every article it produces to countless systematic preparatory tests and studies ... before its standard form is arrived at', said Gropius, 'so the manufacture of standardised building parts demands systematic experimental work'[41]. Schwagenscheidt's work on insolation was thus seen as analogous to the pre-production research of the Ford company: since everyone was going to have the same orientation, it had to be established which orientation was the best. The resultant 'standard form' would be based on the objective findings of scientific research, and not just a few, but all, would enjoy this optimal product. As the corollary of a rationalised production process and a rationally designed product, there was no doubt for these architects that *Zeilenbau* was *the* rational method of site planning.

Unwin, CIAM and modern architectural theory

No one would deny that there were substantial differences in formal and visual terms between the work of Unwin and that of the CIAM architects. Nonetheless it is evident from the above that the thinking from which the forms were produced had a good deal in common. First, and most importantly, both held the view that social welfare, not just aesthetic gratification, was the fundamental goal of architecture. As Ernst May recalled of Unwin, while 'the style of architecture has changed ... the basic outlook ... has not changed but has actually developed further in the direction Sir Raymond initiated': viz. that the 'welfare of men should be the only measure of our endeavours'[42]. Unwin and the CIAM architects also agreed over the main constituents of that welfare: first, a quasi-rural environment of greenery and open spaces; and second, a dwelling to which health-giving sunshine had full and easy access. Furthermore, they agreed over the means by which these desirable results were to be made available to the mass of the population: savings in developmental road construction costs (including the separation of pedestrian from vehicular circulation) and, at least in May's case, satellite planning. Nor did this reflect a belief that purely formal or technical devices could provide the solution to social contradictions: the CIAM architects were as insistent as Unwin that the solution of the economic problems of social housing required political, and not just technical, innovations[43].

The differences between the thinking of CIAM architects and that of Unwin were, however, as important as what they had in common. For Unwin, the 'common-sense' of the architect was a sufficient guide in the step from the interpretation of data to the formation of design[44]. For the CIAM architects this was not the case: for them this could all too easily disguise the imposition of the personal preferences of the architect. In their view, for architecture to be valid it had to be based on the impartial and objective findings of science. In his opening address, May told the second meeting of CIAM (Frankfurt 1929):

> The hundreds of questions cannot be left for their solution to the architect alone, in particular not where, as is being done frequently, he is in the habit of valuing things, under the cloak of economic considerations, from a one-sided aesthetical standpoint and may even want to foist his personal living and dwelling requirements upon the mass.... [45]

The way around this problem lay in the adoption of scientific procedures. Science (in their view, objective and impartial) would provide the data (objective and impartial) on which the designs (objective etc) would be based. May continued:

> Exact observation of the biological and sociological condition of the human beings ... will keep us free from useless theories and lead us to our goal[46].

It was this belief that lay behind the CIAM programme of research. As Catherine Bauer wryly noted, CIAM had a 'certain weakness for expertise: sun-and-air specialists, standardisation specialists, experts in modern spatial aesthetic, and so on'[47]. It was the job of these specialists to provide the scientific basis for the designs. Victor Bourgeois told the delegates at the 1929 CIAM that 'architecture may well be ranked with what is known as the exact sciences', for, like experimental science, it followed 'the analytical method, which proceeds from the investigation of facts to the formulation of rules'[48]. This meant for CIAM that architecture worked, not on the raw data (as for Unwin), but on the general rules and universal truths produced by scientific analysis. Thus the site plan was derived not from the particularities of the individual site, but from the general proposition that optimal insolation was achieved at an aligment of 22.5° from north-south.

It is here that we reach the core of the difference between Unwin and the CIAM architects. For Unwin, the definition of socially responsible design was that it generated an individual solution to each problem: uniformity and standardisation were associated only with the hated capitalist. But for the CIAM architects the position was reversed: this individualising approach of Unwin was exactly what was wrong with architecture and building, a bespoke method of production left over from an antiquated era. For them rationality was so firmly embedded in the production methods of Ford and Taylor that to approach design on the basis of particularity rather than universality could not be other than irrational. Gropius made the analogy with shoes: 'it would no longer occur to 90 per cent of the population to have their shoes specially made to measure. Instead, they buy standard products off-the-peg.'[49] If dwellings were to be made as available to the population as shoes, the same methods of production would have to be employed. From this viewpoint Unwin's method was a leftover from the days of individual, and therefore irrational, production. Ford had replaced the individual client with his particular requirements by the mass consumer with standardised requirements; for the CIAM architects the same transition, from particularity to generality, had to be made for housing to be produced in a rational manner. In other words, both Unwin and CIAM thought of themselves as rational. But whereas for Unwin the rationality of design lay in its derivation from particulars, for the CIAM architects it lay in its derivation from generalised truths that had been arrived at by scientific analysis. It was this belief in abstraction and science, inherited from the Fordisation programme, that above all distinguished the thinking of CIAM from that of Unwin.

This chapter began with a discussion of the versions of the history of modern architecture that have been passed down through the years; the conclusion must be that in fundamental respects these histories misrepresented what actually took place. Not only did they omit the indebtedness of the new architecture to Unwin and the English tradition, but they also passed over crucial elements in the new architecture itself, most importantly, the Fordisation programme on which it was so largely based and from which it drew many of its most important concepts. The reasons for these omissions are to be found, undoubtedly, in the events of the 1930s. Exiled from its country of origin by the Depression and the Nazis, the new architecture was thrown onto the international market in search of a new home and its image had to be tailored accordingly. In the economic and political transformation brought by the slump, social housing lost its importance throughout Europe; with governments cutting back or suspending altogether their housing programmes, it was hardly in the interests of the new architecture to acknowledge openly its basis in the social housing programme of Weimar Germany. In this sense, the relationship of the new architecture to Unwin was part of a much broader one – to social housing in general – that it was better to pass over. Similarly, in the wake of the Wall Street Crash American big business no longer enjoyed the golden reputation of a decade earlier, and furthermore Ford had lost (to General Motors) the aura and dominance in the automobile world that it had enjoyed 20 years previously. It was politic, therefore, to pass lightly over the indebtedness of the new architecture to American production techniques in general and to Fordisation in particular.

In other words, almost as soon as the new architecture was formed, there existed good reasons to tell its story differently from 'how it really was'. What is less explicable is the longevity in this country of the versions of the story which were established at that time. Right up to the present they have exerted a fundamental influence, even where the intention was consciously revisionist. But it is evident that so long as we remain within the terms set by the original historians and interpreters, we will never achieve a real understanding of modern architecture and the thinking on which it was based.

Chapter five

Sellier and Unwin

It is well known that Henri Sellier, the person chiefly responsible for the ring of municipal garden suburbs (*cités-jardins*) built around Paris in the 1920s and 1930s, greatly admired Raymond Unwin, and indeed that he derived a great deal of his thinking about *cités-jardins* from Unwin (Figs 46 and 47)[1]. But what exactly did Sellier learn from Unwin? There were so many phases in Unwin's career, and so many aspects to his thinking, that just to say 'Sellier learned a good deal from Unwin' does not, in itself, get us very far. Was it Unwin the socialist activist? Unwin the arts and crafts architect? The advocate of standardisation? The town planning expert? The pioneer of building research? The regional planner? In particular, the causes with which Unwin was identified when he first established

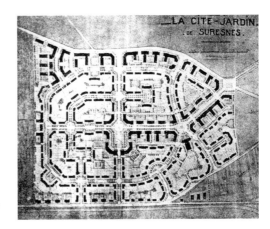

Fig. 46. Office public des habitations à bon marché, La Cité-Jardin de Suresnes, layout, 1919

Fig. 47. Suresnes, paired cottages, 1919–1920

a European reputation before the First World War were rather different from those he espoused after the war and in the 1920s. In the first case, as architect of Hampstead Garden Suburb (1905–) and as author of *Town Planning in Practice* (1909) [CD], his work assumed only a limited intervention by the state into the production of housing: production by private enterprise was still taken as the norm, the question being precisely how to regulate and improve it so as to make its products satisfactory. But in the second case, as principal author of the Tudor Walters Report (1918) [CD] and as chief architect at the Ministry of Health (1919–1928), Unwin was clearly identified with the assumption by the state of responsibility for housing production: the state was not just to regulate private enterprise, as the notion of 'town planning' implied, but actually to replace it in the production of housing.

These remarks indicate that it was possible to 'learn from' Unwin a number of different 'lessons'; or, in other words, to 'read' Unwin in a number of different ways. Here I wish to establish the way in which Unwin was read by Sellier. I shall not be concerned with establishing the extent of Sellier's indebtedness to Unwin; the question whether Sellier's thinking was really as dependent on Unwin as Sellier himself made out is one that will have to be dealt with elsewhere. My aim is merely to establish from which part of Unwin's work it was that Sellier drew when he studied Unwin. To do this I shall focus on the statement on the concept and design of the *cités-jardins* written by Sellier and dated 1 January 1919, entitled 'Le rôle et les méthodes de l'Office public des habitations à bon marché du département de la Seine'[2]. I shall argue that, while Sellier referred repeatedly to, and quoted extensively from, Unwin, the Unwin on which he drew was the Unwin of the early (pre-1914) period, that is, of Hampstead Garden Suburb and *Town Planning in Practice*. This meant that Sellier omitted the changes in Unwin's thinking brought about by the war[3]. Sellier's reliance on *Town Planning in Practice* for his understanding of Unwin meant that Sellier's 1919 policy for *cités-jardins* was, in the European housing context, decidedly old-fashioned.

The design of the *cités-jardins*

In describing the general principles to be followed by the architects of the *Office public* in the design of the *cités-jardins*, Sellier made frequent reference to Unwin, particularly to Hampstead Garden Suburb and *Town Planning in Practice*. There were several drawings and photographs of Hampstead Garden Suburb as well as some of the key diagrams from *Town Planning in Practice*, including the vanishing street perspective illustration (Fig. 48). Unwin's book, which Sellier termed *La Pratique de l'Aménagement des Villes*, was repeatedly cited as the authoritative text of the international garden city movement. For instance:

> As for the general rules to be followed in determining a plan, we can only refer
> to the numerous English and German publications which have disseminated
> the previous experience of these countries, and notably to the remarkable work

by Raymond Unwin that constitutes the synthesis, as complete as possible, of all these publications[4].

Entire sections of Unwin's book, sometimes several pages long, were translated by Sellier and reproduced verbatim. Perhaps the most important of these was in the section of the text where Sellier discussed the methods by which the site plan was prepared. Here Sellier wrote that: 'the principles set out in chapter four of the remarkable work by the eminent English architect, *Town Planning in Practice*, the chapter dealing with the studies to be undertaken prior to the elaboration of a plan, have been exactly applied by the architects of the *Office public*.'[5] He then devoted the next two pages to a translation of the main paragraphs from this chapter, in which Unwin argued that the site plan must derive from a study of site conditions, topography, access and traffic requirements etc, and not from the preconceived idea of the architect (see chapter four). Sellier commented: 'One could not better express the sense of the general method which the architects of the *Office public*, with a common accord, have employed.'[6]

Fig. 48. Henri Sellier, La Crise du Logement, *1921. Sellier's original caption read: 'A successful layout of a crossroads in an English garden suburb: the streets do not meet at a right angle: the angles are "set back". From Unwin, Town planning in practice'*

On other elements of site planning also Sellier quoted Unwin with enthusiasm. He quoted at length the critique made by Unwin of the aesthetic effect produced by detached and semi-detached houses, including Unwin's conclusion that the only answer was 'grouping'. The format developed by Unwin to achieve this – the combination of four dwellings into a single building, with a passageway running through the middle at ground level to provide rear access – was likewise endorsed by Sellier. Similarly on the decoration of roads by planting, Sellier stated, 'Unwin furnishes in this regard a number of suggestions which we believe must be followed'[7] and went on to quote four pages of *Town Planning in Practice*.

On the question of boundary walls, Sellier stated: 'On this point also, we cannot do better than to reproduce the analyses provided by Unwin in the penultimate chapter of his book…. '[8]

Only on one point of design did Sellier actually take exception to Unwin. This was on the question of the relationship between individual variety and overall harmony in design. As regards buildings, Unwin's belief was that 'the variety of each must be dominated by the harmony of the whole'. But for Sellier the danger was not of excessive variety, but of excessive uniformity and monotony, brought about by the use of standard dwelling types. In this context he even ventured a criticism of English practice:

> A fault of some of the English villages built in recent years consists precisely in constructing on [layout] plans – judiciously drawn up and against which no reservation could be expressed – buildings of the same type repeated indefinitely over hundreds and hundreds of metres, creating a most unfortunate impression of uniformity[9].

By this date (1918–19) this was a criticism being made by some of those who had seen the housing schemes designed by Unwin for the Ministry of Munitions, particularly at Gretna (1915–8) (see chapter one). Whether Sellier had visited these schemes himself, or was relying on reports and photographs, is not clear. All this, it should be noted, did not mean that everything about design in *Town Planning in Practice* found its way into Sellier's text, or, conversely, that everything about design in Sellier's text was also to be found in *Town Planning in Practice*. Such was manifestly not the case. The limitation on housing density (the '12 houses to the acre' [0.4 hectare] rule) was fundamental to Unwin but was not mentioned by Sellier; equally the use of apartment blocks, which was important to Sellier, would definitely not have been endorsed by Unwin. The point is simply that, whenever Sellier cited Unwin on the design of the *cités-jardins*, it was the early Unwin of *Town Planning in Practice* that he cited.

Private enterprise versus municipal provision

What about the concept of the *cité-jardins*? What was to be its role in the development process, the mechanics of its construction, its relationship to both private enterprise and the state? Here again we find Sellier drawing extensively on Unwin; and here again it is on the early Unwin of *Town Planning in Practice* that he draws.

Sellier noted that the term *cité-jardins* was confusing and had several different meanings. What he meant by the term, he said, was not the garden city idea per se, as formulated in theory by Ebenezer Howard and put into practice at Letchworth Garden City (1903–). Rather it was the idea that 'it was possible to ameliorate and reform existing towns'[10]. What distinguished schemes such as Hampstead Garden Suburb was their recognition not just of the needs of the individual but also of the community. According to Sellier, 'we perceive in the spirit of the instigators of the garden city movement a sense of social needs, a desire to organise a collective life and to procure for all the enjoyments [hitherto] reserved for a few'[10].

The role of the *Office public* was not to conduct social experiments like Letchworth, but to undertake the realistic task of showing developers that, even within existing economic constraints, it was possible to produce an environment that met basic requirements of health, beauty and social life. The *Office public*, he stated:

> has a limited and clear objective, namely: to build developments of a kind to ensure the decongestion of the City of Paris and its surroundings; to serve as an example to the developers who for the past 30 years have literally sabotaged the suburb; and to show how, while fully taking account of the normal economic and moral conditions of urban life, it is possible to supply the working population, both manual and intellectual, with a dwelling that provides the maximum of material comfort, hygienic conditions of a kind to eliminate the drawbacks of the big cities, and aesthetically pleasing forms of layout that contrast singularly with the hideousness of the methods previously adopted.[11]

With Paris substituted for London, this had been the programme of Hampstead Garden Suburb.

Sellier stated that in achieving this goal, there were two essential requirements: both the increment in land values brought about by development, and aesthetic control over that development, had to remain in the hands of the community – in this case, as represented by the *Office public*. These were two of the essential elements in Howard's garden city idea that had been taken over unchanged in garden suburbs such as Hampstead. The absence in France of a leasehold system, Sellier warned, would make this difficult, but it was absolutely essential that it be achieved. Under the heading 'Necessity of avoiding all speculation and imposing rigorous controls', Sellier wrote that, whoever built the houses, 'their construction must be subject to precise rules from which it is essential not to depart. It is indispensable likewise that the plots are not subject to speculation ....'[12] And again, in summarising the major principles to be followed by the *Office public*, he reiterated the point. It would be the job of the *Office public* to ensure 'the pre-eminent right of the community in that which concerns the recuperation of enhanced values, the elimination of all kinds of individual speculation and the safeguarding of the aesthetic and hygienic rules that are the foundation of its projects.'[13] Such too had been the role of the Hampstead Garden Suburb Trust. Like the Trust, the *Office public* would look to other bodies to build the houses, but would maintain aesthetic control and keep the increment in land values for the community. The only question was how best this was to be achieved.

The *cités-jardins* of the *Office public* would resemble Hampstead Garden Suburb also in that they would cater, not just for the working class, but for all classes of the community. Sellier stated: 'the *cités-jardins*, such as we conceive them, cannot be destined for one limited category of the population.' In this passage Sellier quoted with approval the paragraph from *Town Planning in Practice* in which Unwin described the ideal of the mixed community pursued at Hampstead Garden Suburb:

> There is certainly, says Unwin, no difficulty in mixing houses of different size…. The construction of suburbs inhabited solely by one class of persons is damaging from the social, economic and aesthetic point of view.[14]

The answer, as at Hampstead Garden Suburb, was social mix: 'We must, says Unwin, get the doctor to live amongst his patients'. Such a system also had the advantage that the ground rents paid by the wealthy could be used to subsidise those paid by the poor.

> If one wants to obtain in some areas higher revenues from the plots that can be used to reduce the rents in others, it is necessary to construct the villas with high rents in the most pleasing areas.[15]

Here again, in the concept of an internal subsidy system operating within the boundaries of the *cités-jardins*, Sellier's thinking was in accord with the policy followed at Hampstead Garden Suburb.

The significance of all this can best be grasped, perhaps, by comparing the programme put forward in Sellier's text with that proposed in another document of the same period. Published two months' earlier than Sellier's text and largely written by Unwin, the Tudor Walters Report was, of course, far larger in scope and content; but like the Sellier text, its purpose was to make recommendations on housing policy for the post-war period. Whereas Unwin in 1909 had taken the regulation and improvement of private enterprise as the task, in 1918 he believed that private enterprise was incapable of dealing with the housing problem and that the government, acting through the municipalities, would have to undertake to build houses itself, and it was on this assumption that the Tudor Walters Report was based. As events turned out, not just in Britain, but in Europe more generally, this was more or less what happened (see chapter six); and as a result in the 1920s the Tudor Walters Report came to be widely regarded as the 'bible' of the new state housing.

On questions of design, Sellier's 'Le rôle et les méthodes' resembled the Tudor Walters Report, in so far as it summarised Unwin's earlier work, particularly *Town Planning in Practice* (Fig. 49). But unlike the Tudor Walters Report, its recommendations on policy were also those of *Town Planning in Practice*. Like the Hampstead Garden Suburb Trust, the *Office public* would acquire land and

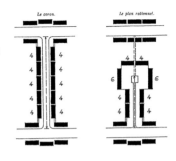

Fig. 49. Henri Sellier, La Crise du Logement, *1921. Diagram from the Tudor Walters Report. Sellier's original caption read: 'Two types of layout: a group of 80 dwellings. In the first example, a layout alas! too frequent, is the terrace in all its ugliness and monotony. The second arrangement, which is much more successful from the aesthetic point of view, is much more economical in terms of road construction, with a road surface three times less than in the first example'.*

lay it out for building, but seek other bodies (either individuals or societies) to undertake the task of building, rather than build all the houses itself. Its role was to regulate and inspire, not to replace, private enterprise in the production of housing. It was a programme closer to the pre-war garden city idea of town planning than to the idea of municipal housing that most countries in western Europe were to follow in the 1920s.

Chapter six

CIAM, Teige and the *Existenzminimum*

Christiane Crasemann Collins and Mark Swenarton

Following its foundation in 1928 at La Sarraz, the *Congrès Internationaux d'Architecture Moderne* (CIAM) held a full conference in October 1929 in the offices of the Frankfurt city architect, Ernst May. At that conference the delegates set themselves the task of investigating, in their various countries, the housing problem of the *Existenzminimum* or subsistence-level population. Their reports were presented the following year at the Brussels conference and collated into a single paper by the Czech critic Karel Teige[1]. Teige's paper was published with the proceedings of the Brussels conference in 1931, *Rationelle Bebauungsweisen*. Written at a time when Europe was reeling under the impact of the Wall Street Crash, it made a critical evaluation of the successes and failures of the European social housing programmes on which many of the architects had been employed – programmes which at that very moment were in the course of being curtailed or abolished. Teige's paper, stemming from investigations undertaken by architects in different European countries, is even today a document of major importance, showing not just the centrality of social housing in the concerns of architectural modernism, but also the determination of the CIAM architects to face up to the realities of the situation in which they were being asked to operate – however stark or unappealing those realities might be.

Social housing in Europe
Although the housing question was attracting increasing attention throughout western Europe prior to the First World War, it was in the years of economic and social turmoil that followed the end of the war that housing became an issue of central political importance. The housing problem created by the war was easy to state but difficult to solve. On the side of housing supply, output had virtually ceased in the belligerent countries long before the Armistice, so that with each year that passed the deficit increased. On the demand side, the end of the war led both to large-scale movements of population, with the troops returning and refugees fleeing from newly-created national states, and to a high rate of household forma-tion consequent on the high marriage rates that followed the Armistice in many countries[2]. Thus the housing shortage, acute in 1918, became worse in the years that followed. In Berlin, the authorities after 1918 ceased recording vacant dwellings and instead recorded the number of households seeking accommodation: by 1922 the number of homeless families had reached 195,000 (Table 1).

Table 1. Number of vacant dwellings and households seeking accommodation in Berlin 1913–1922[3]

Year	Vacant dwellings	Households seeking accommodation
1913	27,811	–
1916	39,863	–
1918	18,972	–
1920	–	72,339
1922	–	194,834

Most European countries (including Britain, France, Germany, Austria, The Netherlands, Belgium) had introduced emergency rent control measures during the war, but as long as there remained a housing shortage on this scale the removal of rent controls was politically impossible. Conversely, as long as rents remained controlled and building costs high, there would be no profits to be made from housebuilding and hence no resumption of large-scale housebuilding by private enterprise. The conclusion, in Germany as much as in Britain, was that houses would be built only if the state intervened[4].

The housing problem was not, however, merely a matter of supply and demand; it was also a political question. The political situation prevailing after the Armistice was unprecedented, with the working class trained in the use of arms and the weakness of the state strikingly demonstrated first by the overthrow of the Russian autocracy in 1917 and then by the collapse of the German and Austrian regimes in November 1918. In Germany and Austria, the revolutions of 1918 at least appeared to put power directly into the hands of the working

Fig. 50. Michael de Klerk, Eigen Haard housing, Amsterdam, third phase, 1917–1920

Fig. 51. Karl Ehn, Karl Marx Hof, Vienna, 1926–1930

class[5], while in Britain the government believed it essential that the state should be seen to meet the demands of the working class for an improved way of life[6]. Either way, the popular demand for decent living conditions was at the top of the agenda. Often socialists and social democrats (such as those controlling the city councils of Amsterdam and Vienna) led the way in proposing new housing initiatives; but, conversely, the political representatives of the employers and the establishment were by no means universally opposed to such measures. Thus the first major post-war housing programme in Europe, the 'homes fit for heroes' campaign of 1919, was launched by a Conservative-Liberal coalition; and although this campaign was terminated prematurely in 1921, a further state housebuilding programme launched by the Labour government in 1924 was retained by the Conservative government after 1925 (see chapter eleven). In The Netherlands, the housing programme of socialist-controlled Amsterdam took place on the basis of legislation introduced by the Liberal national government (Fig. 50)[7]. Only in Vienna, where political divisions reflected the enmity of land-lord and tenant, was there a simple polarisation between the political parties with regard to the housing programme (Fig. 51)[8].

Thus a view across Europe in the mid to late 1920s would have found few countries without social housing programmes of some sort (Table 2). Rarely did governments build directly, almost always preferring to act at one remove, through local authorities or public utility organisations. According to the International Labour Organization (ILO) report of 1930, in most countries (excluding France, for which accurate figures were not available) these bodies accounted for about a quarter of dwellings built since 1918 and in Britain and Germany for over a third. In the non-belligerent Netherlands, the introduction of social housing programmes predated the Armistice and continued until the mid-1920s; in Britain, as already noted, a major programme was launched within weeks of the Armistice. In other countries, however, housing programmes had to wait for the stabilisation of the markets for capital on which house construction depended; the legacy of debt left by the war, plus the subsequent quest for 'reparations', meant that financial instability continued for a number of years and it was not until well into the 1920s that the necessary financial conditions for large-scale housebuilding operations existed.

Table 2. Percentage of total number of dwellings erected by public and private enterprise 1914–1929[9]

Country	Year	Local authorities	Public utility societies	Private enterprise
England and Wales	1919–29	36	–	64*
Holland	1921–29	11	18	71
Austria	1914–28	73	9	18
Germany	1927–29	11	31	58

* including public utility societies

In Austria, stabilisation of the currency was achieved in 1922 by means of loans negotiated through the League of Nations and the following year the socialist council in Vienna announced its programme to build 25,000 dwellings in five years[10]. In Germany, the stabilisation of the currency after the disastrous inflation of 1923 was achieved with the American loans secured under the Dawes Plan of 1924; and a new tax on rents introduced the same year – the *Hauszinssteuer* – opened the way to a major housebuilding programme[11]. Thus, in Frankfurt (one of the best-publicised but by no means the largest of the municipal housebuilders of the 1920s) a programme for 15,000 dwellings to be built over a 10-year period was launched in 1925 (Fig. 52)[12]. In France, the situation was somewhat different, not only because the industrial working class was smaller and weaker than Britain or Germany, but because (partly as a result of the pursuit of reparations) uncertainties in the money markets continued there longer than elsewhere, with currency stabilisation coming only in 1926–1927. This was followed in 1928 by the passage of the *Loi Loucheur*, initiating what was, by European standards, a relatively modest housing programme, targeted primarily to private owners[13].

CIAM architects and the housing programme

The commitment of European governments to housebuilding on this scale opened up a major new field of employment for architects. The housing programmes aimed not merely to relieve the housing shortage, but to effect an improvement in the standard of housing for the working class, and for this the engagement of architects was considered necessary. In Britain the Tudor Walters Report [CD] of 1918 set out at length the argument for an improvement in housing quality and called for every housing scheme to be designed by a qualified architect[14]. In Germany, municipal schemes were designed by architects such as Otto Haesler (Celle), Ernst May (Frankfurt), Martin Wagner and Bruno Taut (Berlin) (Fig. 53), Wilhelm Ripahn (Cologne) and Walter Gropius (Karlsruhe), some of whom built up a large staff in support; at Frankfurt, May's team included Stam, Kramer, Boehm, Kaufmann, Migge, Schütte-Lihotzky, Schwagenscheidt and others. Furthermore, following the British example (see chapter eleven), the German government set up a building and housing research agency, the Reichsforschungs Gesellschaft für Bau und Wohnung (RFG), in 1927, which undertook both theoretical and practical research, including Alexander Klein's housing studies and Walter Gropius's Torten Siedlung in Dessau (Fig. 54)[15].

Thus housing, which formerly had been of marginal importance to architects, now became a central issue to architects and to architecture[16]. In a relatively short period – in Germany, in effect between 1924 and 1930 – a wide range of experiments in the design and construction of housing was carried out and evaluated. In Germany, there was a powerful current of thought, dating back to before 1914, that looked to a 'new start' in architecture, and this received powerful reinforcement from the revolutionary events that followed defeat in 1918[17]. Thinkers of

Fig. 52. Ernst May and others, Römerstadt Siedlung, Frankfurt, 1927–

Fig. 53. Bruno Taut, Hufeisen Siedlung, Berlin, 1925–

Fig. 54. Walter Gropius, Torten Siedlung, Dessau, 1927–28

this persuasion regarded even the name of architecture as old-fashioned and irrelevant. To the architects of the *Neues Bauen* (the New Building), the introduction of the state welfare programmes called for a complete change. Instead of working for wealthy individuals or institutions, architects were now working for the people; instead of aesthetics, it was now social considerations – the 'needs of the people' – that counted. Above all, they believed, the new political goals could be achieved only by an all-encompassing cultural transformation. The task facing the architects was to produce dwellings responsive to the practical and emotional needs of those who traditionally had been voiceless within architecture.

The means to do this, it was believed, was 'rationalisation'. The La Sarraz declaration of CIAM stated:

The need for maximum economic efficiency is the inevitable result of the impoverished state of the general economy.

The most efficient method of production is that which arises from rationalisation and standardisation. Rationalisation and standardisation act directly on working methods both in modern architecture (conception) and in the building industry (realisation).

Rationalisation and standardisation react in a threefold manner:

a) they demand of architecture conceptions leading to simplification of working methods on the site and in the factory;

b) they mean for building firms a reduction in the skilled labour force ...;

c) they expect from the consumer (that is to say the customer who orders the house in which he will live) a revision of his demands.... [18].

Rationalisation referred particularly to the methods of management and production developed by FW Taylor and Henry Ford (see chapter four). Dwellings were to be rethought and redesigned according to the criteria of scientific efficiency, with kitchens and floor-plans that were as efficient as factories. In this process, the traditional notion of the home would be transformed in the light of ideas about communal living and the relationship of women to household labour[19]. The building process was to be reorganised on Taylorist lines, with – as Gropius proposed in 1928 – the architect taking on the role of production engineer[20]. Henry Ford's example showed the way forward. Ideas from the past – about form, aesthetics, sentiment, 'the home' – were now irrelevant: what was needed was the *Wohn-Ford* or 'Ford-dwelling'[21].

When the international leaders of the new architecture got together at La Sarraz in 1928, it soon emerged that there were two rather different views as to what the 'new architecture' comprised. On the one hand was the *Neues Bauen* belief of the Swiss and German architects (Schmidt, Meyer, May) that aesthetics were unimportant compared to the social objectives of economy, function and rationalisation; and on the other was the more traditional view of Le Corbusier that architecture was still ultimately an aesthetic affair, and that what mattered was an architecture appropriate to the idea (rather than the pragmatic requirements) of the new age. Le Corbusier was certainly the most famous, in international terms, of the architects gathered at La Sarraz, but he did not entirely get his way, as the quotation opposite, with its implied equation of architecture with housing, indicated[22]. The two different views of what was involved were enshrined in the different titles adopted for the organisation in the French and German languages, the one positing a new architecture (*Les Congrès Internationaux d'Architecture*

Moderne), the other proclaiming the New Building (*Die Internationalen Kongresse für Neues Bauen*).

The influence of the latter group was clearly evident in the choice of the subject and venue for the 1929 conference. Held in the offices of the Frankfurt city architect Ernst May, the subject was the subsistence-level dwelling (*Die Wohnung für das Existenzminimum*), and an accompanying exhibition displayed approved plans of economical and rationally-arranged dwellings. Official delegates to the congress included May and Häring (Germany), Aalto (Finland), Le Corbusier and Barbe (France), Stam and Rietveld (The Netherlands) and Schmidt and Steiger (Switzerland). Le Corbusier was on a lecture tour in South America and did not attend, although he sent a paper (read by Pierre Jeanneret)[23]. Ernst May's keynote speech identified the subsistence-level dwelling as the key problem for architects. The social housing programmes of the 1920s, May stated, had succeeded only in housing the better-off: to provide houses for those on minimum incomes, subsidies had to be targeted more effectively to those most in need, and architects had to find a rational solution to the problems of the design and construction of the minimum dwelling[24]. It was decided that more information was needed and that the national groups should report back to the next conference on the housing conditions of the subsistence-level population in their various countries. It was these reports, presented to the 1930 Brussels conference on low-, middle- and high-rise building, that Teige undertook to collate and summarise in his paper.

Karel Teige's 1930 overview

Although not an architect, Teige was a versatile and dynamic figure who, as the prime theorist and organiser of the Czech avant-garde, had a considerable international influence during the seminal years of the *Neues Bauen*. Born on 13 December 1900 in Prague (where he was also to die, on 1 October 1951), Teige is best considered as a theoretician of culture: a prolific writer, he was editor of a number of important periodicals, including *Stavda*, and the author of the manifesto *Poetisme* (1924; second version, with Vitezslav Nezval, 1928). He played a leading role in constructivism, surrealism and Dada, both in Czechoslovakia and abroad. A convinced Marxist, Teige rejected any direct or vulgar link between art and politics, searching for an art that had a meaningful function in modern industrial society; his belief was that art should be visionary and reflect the new life of the future. In the 1920s, Teige had published several articles on architecture, and had lectured at the Bauhaus at Dessau on architecture and sociology. He had not attended the inaugural meeting of CIAM at La Sarraz in 1928 but participated at the subsequent two meetings, as delegate of the Czech group which was almost identical with the *lèva fronta* (Left Front). His views on architecture, planning and housing sprang from the constructivist and functionalist theories to which he adhered throughout his life[25].

Teige's paper evidenced the disillusionment felt by the architects at what they increasingly saw as the failures of the social housing programme. Drawing on information from the countries of mainland Europe and the USA, it reiterated the point already made by May at the 1929 conference: that the state housing programmes of the 1920s had been 'of much greater benefit to the so-called middle class than to the real subsistence level, because for the low-income classes even the subsidised housing remained out of reach'[26]. This was the constant point of attack made by the Communist Party on the social housing programmes in Germany and by the end of the 1920s the accuracy of the charge appeared undeniable: at some of the Frankfurt *Siedlungen*, such as Praunheim and Römerstadt in the Nidda Valley, the lowest rents were equal to half of the wages of unskilled workers, with the result that the housing was largely occupied by skilled workers, clerks and officials[27]. The problem faced by all the housing programmes (with the notable exception of Vienna) was that rents had to include the amortisation of, and interest charges on, the capital borrowed to pay for initial construction, and so even when subsidised the rents were too high for unskilled workers[28].

Teige followed up this observation with two further points closely related to the Communist Party's position. First, he pointed out that the housebuilding programme was subsidised not just out of the yield of taxes on rents (in Vienna, the *Wohnbausteuer* so prominently advertised on the Viennese housing blocks), but also by indirect taxation which fell most heavily on the poor, so that often the poor paid for the housing of the better-off. Secondly, he related the inability of the housing programme to house the poor not to some temporary or accidental defect, but to the inherent structure of capitalist society: the 'social structure of today's population ... provides the answer as to the origin of the current housing shortage'[29]. In arguing that the housing problem was a necessary part of capitalist society and would disappear only with the disappearance of that society, Teige was following in the tradition established by Engels in his celebrated study of *The Housing Question*.

When the Frankfurt conference decided to investigate subsistence-level housing conditions, the housing programmes of most European countries were intact. Even while the conference was taking place, however, an event occurred on the other side of the Atlantic – the Wall Street Crash – that within a year or two was to bring chaos to the European economies and the demise of the social housing programmes. As economic output fell and unemployment rose (in Germany from two million in 1929 to five million in 1931), increasingly right-wing governments cut back their social housing programmes in the interests of both private enterprise and public economy. As Teige noted, the 'restoration of private housebuilding and the free market are the dominant tendencies in nearly every country'; in 1931, the Hindenburg government axed the housing programme in Germany and over the next two years the same happened in Britain, Austria and France[30].

For the CIAM architects, it was events in Germany that had the most serious effect, for it was on the housing programme there that many of them relied for their employment. Even before the 1930 Brussels conference Ernst May and many of his team from Frankfurt (including Mart Stam and Hans Schmidt) had left Germany for the Soviet Union; they were followed soon after by Hannes Meyer and a little later by Bruno Taut[31]. The exodus intensified with the Nazi takeover in 1933, although by then the Soviet Union was no longer so amenable to CIAM architects; Gropius left Germany in 1934 for Britain and the USA. With the loss of its German core, the future of CIAM was thrown into doubt: no conferences were held in 1931 or 1932, and when the next conference took place in 1933, it was evident that CIAM had changed significantly. The personnel was different; Le Corbusier was indisputably pre-eminent; and the location – on a luxury steamer in the Mediterranean – was symbolic of the period of exile into which modernism had been driven by the events of the early 1930s[32].

Teige's paper, then, belongs to that first phase of CIAM, in which the architects attempted to come to terms with the wider forces within which they operated. The CIAM architects of 1929–1930 believed that if their efforts were to be directed at the problem of subsistence-level housing, they should first analyse the economic and material conditions that made subsistence-level housing an intractable problem irrespective of national boundaries. Hence the enquiry undertaken by CIAM architects of the various countries on which Teige's paper was based. The conviction that architecture should proceed only on the basis of such understanding was driven into eclipse by European events of the 1930s, but with the global take-up of modernism after the Second World War was to have a powerful effect, in Europe and the developing world alike.

Chapter seven

The education of an urbanist

The notion that Ruskin and Morris provided Raymond Unwin (1863–1940) with much of his intellectual inspiration derived from pronouncements made by Unwin himself in his later years. In his inaugural presidential address to the Royal Institute of British Architects in 1931 Unwin stated that:

> my early days were influenced by the musical voice of John Ruskin, vainly striving to stem the flood of a materialism which seemed to be overwhelming the arts, and much else; and later by the more robust and constructive personality of William Morris, and his crusade for the restoration of beauty to daily life.[1]

On other occasions in the 1930s Unwin recalled that as a schoolboy in Oxford in the 1870s he had seen Ruskin leading the road-diggers at Hinksey, and that while working in Manchester the following decade he had been inspired by William Morris to set up a local branch of the Socialist League. On receiving the RIBA Gold Medal in 1937 Unwin again spoke of the formative influence of Ruskin and Morris, although he also mentioned alongside them James Hinton and Edward Carpenter:

> One who was privileged to hear the beautiful voice of John Ruskin declaiming against the disorder and degradation resulting from laissez-faire theories of life; to know William Morris and his work; and to imbibe in his impressionable years the thoughts and writings of men like James Hinton and Edward Carpenter, could hardly fail to follow after the ideals of a more ordered form of society, and a better planned environment for it, than that which he saw around him in the 'seventies and 'eighties of [the] last century.[2]

The notion that Ruskin and Morris were fundamental to Unwin's intellectual development raises two questions. The first relates to the socialist strategy informing Unwin's architectural thought. In the 1880s Unwin fully endorsed the argument of Morris and the Socialist League that socialists should avoid any entanglement with the electoral process and the state. Yet his thinking as an architect and town planner as set out from 1901 onwards was predicated on precisely the opposite point of view – on using the state to further socialist reform, particularly in the area of housing and town planning. How did Unwin come to make this change from the 'abstentionism' of the Socialist League to the 'municipal socialism' of the 1900s?

The second question involves Unwin's relationship to one of the central tenets of the tradition derived from Ruskin's characterisation of 'The Nature of Gothic', the concern with architecture as the product of free labour. Although Unwin periodically made obeisance to the notion of architecture as the expression of joy in labour, in practice his thinking was concerned with the role of architecture in giving satisfaction, not to its producers, but to its users. Unwin's interest was in the role that architects and architecture could play in the process of transition to a socialist way of living, whether at the level of the individual, the community or the city. How did Unwin come to differ from the rest of the English Ruskinian tradition in this respect and think about architecture from the point of view not of the builders but of the occupants?

The answer to both these questions is to be found in Unwin's position as an intellectual active in the socialist movement in the north of England in the 1880s and 1890s. Here the decisive intellectual influence on Unwin was not Ruskin or Morris, but Edward Carpenter (1844–1929), the socialist, poet, erstwhile clergyman, advocate of sexual (particularly homosexual) liberation and propagandist of 'the simple life'. But even more than any particular intellectual influences, Unwin's development was moulded by the dynamics of the socialist movement in the industrial north in these decades. From this position in the northern socialist movement Unwin was led away from both the concern with joy in labour and the abstentionist politics that appealed so strongly to the socialist intelligentsia in London and the south, towards the application of his architectural skills to the attempts being made to produce 'practical socialism', particularly on the basis of the municipality.

In the autumn of 1901 Unwin was transformed from an obscure provincial architect into a figure with a national reputation. Within the space of a few weeks *The Art of Building a Home* (written with his partner Barry Parker) [CD] was published, gaining wide attention, and Unwin gave widely reported lectures to three national bodies, the Garden City Association, the Workmen's National Housing Council and the Fabian Society, this last being published under the title *Cottage Plans and Common Sense* (1902) [CD]. The ideas on housing and town planning that Unwin set out here and in other publications over the next few years have come to be seen as a turning-point in the relationship between archi-tecture and the city, with major effects not only in Britain but throughout the world[3]. Here I want to trace the development of Unwin's thought and see how he came to formulate these ideas.

The impact of Edward Carpenter

In later life Unwin recalled that he 'had grown up among liberal ideas in religion and politics'[4]. In particular he took from his family a marked antipathy to the ethics of commerce. When Unwin was 'ten or eleven years old', his father sold the family business in Rotherham, West Yorkshire, in uncertain circumstances and moved with his family to Oxford, to become a student at Balliol College, matriculating in 1875[5]. Unwin's father alerted his son to the social questions of the day (together they attended the first meeting of the Land Nationalisation Society in 1881) and inspired him with a hatred of business. Several years later Unwin quoted his father's comment that a limited liability company was 'a company with unlimited ability to lie'[6]. In his first published article, published in July 1886 in Morris's *Commonweal*, Unwin asked:

> What is likely to be the result on the character of a race of men if they are set to compete with one another, each to get the better of his neighbour? Surely they must become selfish and heartless. The most selfish will get on best…. Where would the modern business man be who, when selling out shares which he believes will go down, should stop to think of the ruin he may bring to some poor family? Again, has it not become a bye-word that a certain amount of dishonesty is necessary in all trades?[7]

This picture of the inherently corrupting effects of the competitive system sprang from family lore as much as from political analysis.

On leaving Magdalen College School, Oxford, in 1880, Unwin's initial intention was to enter the church. But he was not sufficiently sure of his vocation to proceed with the expense of a university education, and instead he went back to the north, to Chesterfield, where he served as an apprentice engineering draughtsman/fitter, and continued to debate his future. At Chesterfield he lived for a time with his wealthy cousins, the Parkers; Robert Parker was a banker, who disapproved of Unwin both for his penury and his socialism, all the more so when Unwin and his daughter Ethel Parker fell in love.

In Chesterfield Unwin encountered Edward Carpenter. Carpenter's conviction that in 'the simple life' he had found the way to reconcile his belief in socialism with his economic position as a *rentier*, was to be of profound importance to Unwin's entire thought. At this stage Carpenter was in the process of abandoning lecturing to pursue the simple life in the country – a pursuit that took him in 1880 via the Totley farm of Ruskin's Guild of St George to live in a cottage nearby with his friend Albert Fearnehough and his wife. Unwin, who attended Carpenter's lectures in Chesterfield in 1880–81, recalled in a tribute to Carpenter:

> From about the year 1881 Edward Carpenter became a great influence in my life. He was then giving one of the last of his courses of University Extension Lectures on Science…. A year or two later, when intercourse with working people and close contact with their lives brought home to me the contrast with all that I had been used to in my Oxford home, I turned again to Edward Carpenter for help, as the overwhelming complexity and urgency of the social problem came upon me.[8]

In 1883 Carpenter acquired some land in the country outside Sheffield, at Millthorpe, and built a cottage where with the Fearnehoughs he set up as a market gardener (Fig. 55). In a letter to Ethel Parker in May 1884 Unwin described his first visit to Millthorpe, in terms that uncannily forecast his future conception of the ideal physical setting of the home:

> I had a little note from Mr Carpenter saying he is now at Millthorpe & asking me to go and see him yesterday. So I walked over and found that he has built a little house just beyond that old tannery…. He has the next three fields on the same side…. Oh it was so beautiful!….
>
> The house he has built is a long one only one room deep, as all the rooms face South, and look over to a beautiful ford. One field is laid out in oats for the horse and wheat for fowl use, the other is in grass with a few young apple trees in it, the centre one in front of the house is planted with fruit, vegetables and flowers – lots of young rose trees – there is a stream running at the bottom where primroses grow.
>
> He is living with a man who was a scyth maker in Sheffield, his wife and daughter, they all share alike the living room where cooking is done and there is a piano in the same room, they seem a very happy family.[9]

Fig. 55. Edward Carpenter and friends outside his cottage at Millthorpe in the 1890s

From then on Carpenter was the major influence on Unwin's intellectual development. He and Unwin were in close personal contact from 1884, when Unwin 'spent helpful and happy weekends with him and his companions' at Millthorpe, with 'long days and nights of talk'[10]. Unwin lived within walking distance of Millthorpe until October 1884 and again from May 1887 until 1896. Even Unwin's time in Manchester (early 1885 to the end of 1886) did not interrupt the relationship, for Carpenter came several times to Manchester to lecture for Unwin's Socialist League branch and Unwin continued to go over to Millthorpe for weekends. For Carpenter doubtless the attraction of the relationship was in part sexual; Unwin seems to have managed to remain unaware of this, even though they slept together. Carpenter figured prominently in Unwin's letters to Ethel Parker, and when in 1891 Unwin was urging her to deepen her knowledge of socialism Carpenter was the first author he recommended[11]. When the Unwins' first child was born in 1894, he was named Edward and Carpenter was godfather.

In October 1884, when Unwin was leaving Chesterfield, Carpenter gave him a copy of *Towards Democracy*, the long poem recounting Carpenter's search for spiritual and emotional freedom that had been published the year before. Despite the personal contact he had had with Carpenter, the effect of the book on Unwin was overwhelming: it was a 'bewildering revelation'[12] that 'opened [the] door to [a] new world'[13]. In 1931 Unwin recalled that:

> The feelings compounded of mystification, escape and joy with which I read it through on the journey to Oxford, are still a vivid memory … the sense of escape from an intolerable sheath of unreality and social superstition which the first reading of *Towards Democracy* brought to me is still fresh.…

The message that Unwin took from Carpenter's poem centred on the need for inner spiritual reform as the basis for new relations both between people and between people and nature. Unwin wrote:

> It is difficult to convey any intelligible idea of such a poem, ranging as it does over all things in heaven and earth – and hell; and depending as it does for its main influence on the artistry of expression, the turn of a phrase, the intimacy of a touch. There are, however, two or three main conceptions towards the expression of which all tends, and every touch is made to contribute. They begin with a new understanding, relation and unity to be realised between the spirit of man and his body, the animal man no longer a beast to be ridden, but an equal friend to be loved, cherished, and inspired.

> Growing out of this, made possible by it, there then emerges a new sense of equality and freedom in all human intercourse and relationships.

> Content, in happy unity with its body, the soul of man, thus accepting equality of spiritual status, and enjoying free communion with its fellows, discovers a new relation to the universe, to nature, and to the Great Spirit which pervades it; a new faith, not of belief in this or that, but of trust.[14]

On another occasion Unwin described *Towards Democracy* as the story of 'a soul's slow disentanglement from [the] sheath of custom & convention'[15]. The great theme towards which Carpenter was working was the 'simplification of life', which he set out in a paper of that name in 1886. The fundamental antithesis posed by Carpenter was between 'convention' (artificial, unnatural, unhealthy and unnecessary) and the 'real needs' of life (simple, natural, open and non-exploitative). Accepting both Marx's theory of surplus value (which for Carpenter rendered dividends morally soiled) and Ruskin's proscription of usury, Carpenter argued that the only course for right-thinking people of wealth such as himself was to abandon the extravagant lifestyle to which they were accustomed in favour of the simple life; by reducing their requirements to, for example, a single dish for each meal, they could minimise the expenditure of labour and wealth required to support their daily lives, using the now super-fluous income from their dividends for socially progressive causes, such as promoting socialism.

These ideas became basic to Unwin's thought, both about socialism and later about architecture. In 1887 he wrote to Ethel Parker:

> sometimes there comes across me a sort of misty idea of some society in which there shall be a 'better way' ... a better land altogether where life would be freer & happier, more natural, everything made pure & clean, clean food, clean lives, clean bodies, & all open & above board. Of course it is the idea of Towards Democracy. Only dear Ettie don't you know how at times it comes over one with more power & seems for ever new....[16]

Carpenter, according to his own account, took the 'style and moral bias' of his socialism from Ruskin and the 'economics' from Marx. By the latter, he meant chiefly the theory of surplus value, extracted from Marx and popularised by Hyndman in *England for All*; when Carpenter read this work in 1883 he found in the chapter on surplus value the essential basis he needed to get 'the mass of floating impressions, sentiments, ideals, etc in my mind ... into shape'[17]. For Unwin there was nothing new in the Ruskinian element in Carpenter; he was familiar with Ruskin's teaching from Oxford and went on to read several of Ruskin's works after meeting Carpenter (including *Modern Painters* and *Unto this Last*), but without any particular expression of enthusiasm[18].

The Marxian theory, by contrast, was unfamiliar to Unwin. Carpenter introduced Unwin to Hyndman's writings (on that first visit to Millthorpe Unwin went away with one of Hyndman's books) and in 1885, as part of his programme of self-education in socialist theory, Unwin decided that he needed to gain a 'better understanding of its scientific basis as given by Marx'[19]. The result was a lecture, 'The Dawn of a Happier Day' (January 1886), which set out at some length what Unwin explicitly called 'the theory of surplus value'[20].

This early grounding in the value theory of Marx sets Unwin apart from Morris and other Ruskinians. But it is important to note that, with Unwin as with Carpenter, acceptance of the value theory did not imply acceptance or even understanding of historical materialism as a whole. Carpenter incorporated the value theory of Marx within an ethical, transcendental socialism; his view of history showed no debt to Marx, being biological and evolutionary rather than materialist, and his concept of art and architecture, as communication between human souls, was Ruskinian and transcendentalist. Unwin followed Carpenter in the idealist notion of both socialism and art: the change to socialism had to start, he believed, with the spread of 'the spirit of socialism'; and he endorsed the Emersonian definition of art, as the 'manifestation afresh of the universal mind or Soul which is behind all things'[21].

Unwin and the northern socialist movement

From 1885 until the early 1890s, the main focus of Unwin's interests and energy was the northern socialist movement, first in Manchester and then across the Pennines in the adjacent towns of Chesterfield, Sheffield and Clay Cross. In the course of this work as a socialist speaker and organiser, Unwin was led away from the abstentionist politics espoused by Morris and the London socialists towards the programme of immediate and pragmatic reform developed by the northern socialists during these years.

Unwin left Chesterfield in October 1884. After a few months at home in Oxford he returned north, early in 1885, to Manchester, close to Ethel Parker at Buxton. In Manchester he had obtained a job as an engineering draughtsman, but he regarded as his real work the voluntary activities he undertook for various social causes. These included the temperance movement, the Ancoats Brotherhood (a social mission with which Morris was connected) and above all the Socialist League, which was established by Morris as a secession from the Social Democratic Federation (SDF) in 1885. In retrospect Unwin came to regard this period as the heyday of his political activism. He recalled in 1902:

> During the lifetime of the Socialist League I found a sphere of work so congenial that I descended for a time into the arena of actual struggle…. Times have changed since the League days. Many who know our movement now would find it difficult to realise the frame of mind in which we worked in '85 or '86, when the coming Revolution loomed large in our imaginations….[22]

The early stages of Unwin's attachment to the Socialist League are unclear. Carpenter's lead was initially ambiguous: personally he supported Morris in forming the League, but Carpenter also had loyalties to Hyndman and the SDF and it was not until September 1885 that he joined the League. Unwin moved a good deal faster. At the end of January 1885 Unwin wrote to Ethel Parker:

> I see there has been a split in the Socialist party. I think I told you Morris and several more have left the Social democratic Federation and have started a 'Socialist League'. I see they are going to issue the first no. of their organ the 'Commonweal' tomorrow, I have ordered a copy to see what it is like. Morris is going to manage it himself so I hope it will be a good paper. May I send you a copy....[23]

A visit to Manchester by Morris in July 1885 won Unwin for the League. The following month Unwin organised the take-over of the local Socialist Union and became secretary of the new Manchester branch of the Socialist League[24]. From then he was indefatigable as organiser and, along with Morris, Carpenter and others, as speaker at the meetings held both indoors and in the open air.

But things did not go well for the League in Manchester: during the winter of 1885–86 membership of the branch fell steadily (from 41 in October to 28 in February) and by March Unwin reported that it was 'not easy for me to say how we stand now that our meetings are so irregular'[25]. Returning in June 1886 from the League's national conference (at which he made friends with May Morris) he found things no better:

> I am afraid there is little to report while I have been away, only one open air meeting was held on two Sundays. Last Thursday 3 turned up to hear my report of conference though I sent all P.cards. It is very disheartening.[26]

In an attempt to revive its fortunes the League branch decided to move, to Openshaw in the north-east of the city, where Unwin and another branch officer took a house, but to little effect. At the end of September Morris paid a visit to Manchester in a vain attempt to restore morale. In November, *Commonweal* reported that Unwin had resigned as branch secretary on leaving the district, and the branch faded away thereafter[27].

The failure of his socialist activity in Manchester weighed heavily with Unwin; it was, he felt, his 'want of enthusiasm' that 'made my work in Manchester come to so little'[28]. A more likely cause was the League's policy of 'abstention' from parliamentary elections, for, as Unwin told League headquarters in June 1886, 'most of our old members are not much good, and the two or three that are have gone head first into the election!'[29]. In this regard it is significant that while the League floundered in Manchester the SDF, which was not committed to the anti-parliamentary line, prospered.

At this stage Unwin adhered unequivocally to the League's abstentionist line; 'personally I am strongly in favour of the League keeping non-parliamentary', he wrote in July 1886,[30] a view that he argued with vigour in an article in the

Commonweal the following month. Reforms aiming to alleviate the condition of the working class within the framework of the existing system were dismissed altogether as a deliberate means of 'postponing any attempt to get at the real cause'. The only answer was to 'revolutionise society':

> I use the word revolutionise, because nothing short of a revolution will do. We have got to a stage where mere reforms are useless, often worse. If you have a good system founded on rightness and harmony, it can be improved by reforms; but where the system is bad ... there is no place for reform....[31]

After leaving Manchester in November 1886 Unwin returned for a few months to Oxford. In May 1887 he started a new job as a draughtsman for the Staveley Coal and Iron Company at Barrow Hill, near Chesterfield, where his work included the design of machinery for the pits. The head of the firm, Charles Markham, was a socialist sympathiser and poet who had contributed to *Commonweal*, but even so Unwin had to be careful not to mix work and politics directly. Instead he involved himself in socialist activity in three nearby towns: Chesterfield, Sheffield and Clay Cross.

In Chesterfield, his friend Joe Cree told him as soon as he arrived in May 1887, the prospects for socialism were not rosy and a reading or discussion group was the most that they could aim at. The group was instituted, meeting on Sundays, and was kept going by Unwin until 1891.[32] Speakers included, as well as Carpenter and Morris, the Christian Socialist John Furniss, from the nearby Moorhays commune, and Unwin himself, on the ethical teaching of James Hinton. Unwin found a strong secularist feeling among the Chesterfield socialists and warned them 'that they must be careful in condemning the ordinary orthodox Xtianity not to imagine they had said anything against what Christ really taught'.[33]

The Sheffield Socialist Society was a larger venture. It had originated during the general election of November 1885, when Furniss, Carpenter and others had put up an independent parliamentary candidate against the Liberals. At the end of February 1886 Morris visited the Sheffield socialists with the hope of gaining their adherence to the League, but found them resistant, mainly because of 'our repudiation of a Parliamentary method, the reasons for which I did my best to explain'[34]. Carpenter provided a programme for the Sheffield Socialist Society, as they called themselves, which included among its immediate objectives 'Labour Representation ... in all forms – Parliamentary, Town Councils, Boards of Guardians, School Boards, etc'[35].

While living at Manchester, Unwin travelled over to lecture to the Sheffield socialists and on his return to Chesterfield in May 1887 he became a regular member of the Sheffield group, lecturing for them frequently. It was as such that Carpenter described Unwin in his autobiography: 'Raymond Unwin, who would come over from Chesterfield to help us, a young man of cultured antecedents, of first-rate ability and good sense, healthy, democratic and vegetarian'[36]. In Sheffield Unwin met socialists from other parts of the country who were making the pilgrimage to Millthorpe, including in September 1887, the ethical socialist Percival Chubb, of the Fellowship of the New Life, whose ideas on ethical education appealed to him.

Unwin remained involved in the Sheffield group until 1890. At this date, after the collapse of the League but before the emergence of the Independent Labour Party, the anarchists in Sheffield and elsewhere were able to gain a hearing for their policy of immediate revolution. In Sheffield the anarchist group included the so-called 'Dr' Creaghe, who, it was stated, would be 'helping us to get the Rev. over speedily'[37]; a 'most revolutionary beggar' was Unwin's comment after having brought Creaghe over to lecture at Chesterfield in January 1891. Unwin listened to the anarchist argument in favour of immediate revolution but was not convinced that 'blowing up bridges' was really the way to bring about socialism[38]. This was perhaps fortunate, since two of the Sheffield anarchists were soon afterwards arrested in connection with the 'Walsall Bomb Plot'.

The third of Unwin's socialist fora, Clay Cross, brought him into direct contact with the parliamentary strategy in the person of the young socialist agitator JL Mahon. With the encouragement of Engels, Mahon sought to bring socialism and the labour movement together by 'taking hold of the working-class movement as it exists at present, and gently and gradually moulding it into a Socialist shape'[39]. Basing himself in the northern coalfields (particularly Northumberland, where a bitter strike was in progress) Mahon set up the North of England Socialist Federation, with a clear commitment to parliamentary action.

Unwin acted as behind-the-scenes organiser for Mahon in the Derbyshire coalfield. In June 1887 Mahon asked Unwin to help organise a socialist rally of the Derbyshire miners at Clay Cross, to which Unwin agreed, although worried at the danger 'if it gets known to our boss that I am stirring up the miners!'.[40] *Commonweal* reported that on 12 July 1887:

> A large meeting of the Derbyshire miners was held. Raymond Unwin [and others] … spoke. Mahon explained the lines on which the Socialist organisation of the Northumberland miners had been formed, and sixty names were at once given to form a similar society.[41]

Other meetings followed; on 2 August Unwin 'spoke for some time on the programme of the North of England Socialist Federation which we have adopted'[42].

The following week Unwin wrote to the Socialist League headquarters on behalf of the Clay Cross socialists to enquire about the conditions under which the League would accept them as an affiliated society, and received the uncompromising reply that if they wished to affiliate to the League they must sign a pledge renouncing parliamentary action. A month later Unwin informed the League that:

[at] a meeting held at Clay Cross last Tuesday your letter was read & the matter of joining the League was discussed. It was decided not to join the League, the majority not being willing to sign any pledge on the parliamentary question. I must say that we regret that the League has decided to push this question in a more extreme manner than formerly.

It was not that they wanted immediate parliamentary action:

in fact the majority of Clay Cross are of opinion that Education and organisation outside all parliamentary work is the best course to follow at present, & would have been glad to join the League as the best organisation for carrying out that work. But they do not feel free to sign anything which might be construed into a repudiation of parliamentary work in the future....

Our position shortly is this. We believe parliamentary action for the present & for some time to come to be useless or worse but we think that eventually the changes may be made through the means of Parliament.[43]

Under the immediate impact of these events Unwin's thinking about the merits of the revolutionary and the parliamentary routes to socialism changed. An article written early in September 1887 set out his change of view. Instead of arguing the abstentionist case against the parliamentary, Unwin now sought for 'a third course', which without being a compromise would take account of the strengths and weaknesses of both sides of the argument. The objection to revolutionary action was that it had no popular support; the objection to parliamentary action was that it was 'likely to lead to no good'. Instead, he believed, a third course was offered by education; this was really the old Socialist League line, but now given an even stronger ethical emphasis. The need was to address the people, 'teach them Socialism, and try by any means in our power to spread also the true Socialist sentiments of brotherhood, freedom, and equality'; because spiritual had to precede social change: 'There must be more of the spirit of Socialism, more regard for others' good, before any social changes can be of much use.'[44]

This remained Unwin's position until the rise of the Independent Labour Party transformed the political landscape in the 1890s: revolution was futile; parliamentary action was unlikely to yield any result; the 'great work' was to educate the people in socialism[45]. Unwin thus deferred to some indefinite point in the future the likely date of the advent of socialism. In August 1887 he noted that his youthful belief in the imminence of socialism was passing: 'somehow each time I come to seriously think of it I seem to see that there will be longer and longer to wait. I used to think of 3 or 4 years, now I think of 7 to 10 or even 20 years!'[46] The problem, as Unwin had found when addressing the Derbyshire miners, was that it was hard to get them 'to work for anything distant, they want something at once'[47].

In the meantime there was one form of immediate action that Unwin felt was legitimate: the founding of socialist colonies or communes. In the area around Sheffield and Chesterfield this was a strong tradition, with the St George's Guild farm at Totley, Carpenter's Millthorpe (which acted as a sort of training ground for other would-be colonists), and the Christian Communist settlement at Moorhays led by John Furniss.

On his first visit to Millthorpe Unwin had wondered 'whether a place something like Mr C.'s would not be a good thing ... a little centre of socialism and refinement for some country place'[48]; and even while in Manchester he had written an article for *Commonweal* commending these 'social experiments'. His return to Chesterfield in 1887 strengthened this interest:

> I have been thinking that Edward is right in trying to do something in a small way even now as well as agitating for a large change.... I wish I could do more in that sort of work.[49]

Unwin visited, and was strongly impressed by, John Furniss's commune at Moorhays, some four miles [6.4 kilometres] from Millthorpe, where 120 acres [48.5 hectares] had been taken over by a group of 20 settlers:

> The communists seem to be doing pretty well.... We dined off a regular communist apple dumplin, the finest I ever saw! There is something very interesting in these fellows, uneducated simple fellows as we should call them, living together a life so much more noble than most of us are able to do.[50]

Furniss impressed Unwin, not only by his socialist work in Sheffield and by his commune, but also by his primitive Christianity. At this stage Unwin was almost as much concerned with questions of ethics and religion as with politics. His diary entry for 22 May 1887 is characteristic:

> Joe [Cree] & I get on very well, we talked of social matters and religious, Sunday keeping and worshipping God by serving man.

> I can't quite make up my mind about morality, whether we are to start with the good of others as the basis & take moral laws as merely the expression of what wise men have thought to be for the good of others, or are we to take them as moral laws given from heaven to be obeyed in spite of all appearances because they are from God....[51]

Unwin's major intellectual interest at this time was in the ethical writings of James Hinton (1822–1875), an interest that he shared with Carpenter. In Hinton, with his belief in the service of humanity and his insistence on the primacy of the moral or spiritual faculty over the intellectual, Unwin found a confirmation of his own beliefs. Unwin even felt that it might be his vocation to proselytise for Hinton. Hence his delight in finding in the uneducated Furniss a philosophy similar to that of Hinton:

> You know I said I wanted to translate Hinton to the people, well there's Furniss, he's hardly read anything but the bible and he has grasped all the vital parts of Hinton.... He thinks nothing can be done unless we have that real religion, we must all be 'saved', that is, from selfishness.[52]

It was not surprising, given this, that Unwin was repelled by the positivist philosophy of Auguste Comte, which he studied at this time but found both anti-spiritual and anti-democratic[53]. Unwin preferred spiritual fare: Emerson, Tolstoy or the *Bhagavad Gita*.

Although none of the three socialist groups with which Unwin was involved between 1887 and 1890 were branches of the Socialist League, Unwin's personal loyalties remained with Morris and the League. After leaving Manchester, Unwin became an individual member of the League and he continued to write regularly for the *Commonweal*. But with the collapse of the League at the end of 1890 – a part of the general disarray of the socialist movement – Unwin found his socialist commitment being put to the test.

Like many others he found it hard to maintain socialist enthusiasm when the only prospect was an indefinite period of the rather thankless work of 'education'. At the beginning of 1891, after a 'rather poor' lecture and meeting at Chesterfield (his only remaining arena of socialist activity), he reaffirmed his belief in socialism: 'the ideal of Socialism is to some of us a religion, & it does Ettie not only tell us what is right but helps us to do it.'[54] But he admitted that he was finding the work of organising socialist meetings increasingly tedious. A few months later, while still affirming that 'the Socialists hold the ideas which seem to me right'[55] he admitted that socialism no longer filled the role of a religion in his life:

> You see Ettie at one time I was sort of given up to Socialism, it was my religion and I feel the loss of it as such. But I think it quite possible for some other side of the work besides the agitating to take the same place.[56]

As a result, as he later put it, in the early 1890s he largely 'yielded to ... idle and cowardly impulses, and retired to my peak'[57]. In these years Unwin was largely preoccupied with personal matters, in particular with advancing his career prospects sufficiently for Robert Parker to drop his objection to Unwin and Ethel Parker becoming engaged. In 1891 agreement of a sort to an engagement was obtained, but then retracted; marriage eventually followed in 1893, whereupon the Unwins set up home in Chesterfield.

The architecture of social experiments

A new period in Unwin's life began in 1896, when he went into partnership with his brother-in-law Barry Parker in architectural practice at Buxton, on the west side of the Peaks. At the same time the Unwins moved from Chesterfield to Chapel-en-le Frith (close to the moors but also on the main route from Manchester to Sheffield), where they occupied a lodge house that remained their home until they moved to Letchworth Garden City in 1904.

Setting up practice meant for Unwin not just a new kind of employment but also release from the constraints so long imposed on his political activities by his employment in industry. This freedom became apparent in January 1897 when for the first time Unwin's name appeared in the list of lecturers published by the *Labour Annual*[58]. More importantly it gave Unwin for the first time the opportunity to combine his skills as a designer with his interest in socialist ventures.

Throughout this period Unwin retained his belief in the ethical mission of socialism. His activities in this regard centred on the Labour Church, of which he was an active member in the late 1890s, lecturing regularly at Labour Churches in the midlands and the north[59]. In an article for the *Labour Prophet and Labour Church Record* in March 1898 Unwin stated that the name 'Labour Church':

> directs attention at once to the two great ideas which have caused the movement to spring into existence. The first of these may be expressed in this way: that the Labour Movement is powerless to move society sufficiently deeply to realise its aims, without the help of religion.... The other idea which is somewhat the converse of this may be put thus: A religion which stands aside from any great movement of social regeneration must cease to be vital and shrivel into mere formalism.

Echoing both Carpenter and Hinton, Unwin wrote that the religion of the labour movement was based not on mere doctrine or intellect, but on something deeper, the religious feelings inspired in the 'simple soul' by 'the words of love, and the life of goodness' of Christ. It was the job of the Labour Church to unite the labour movement around:

> the broad ethical principles and ideal feelings which are common to us all, putting these forward as the main basis of our movement. Those merely intellectual beliefs about economic laws, or political tactics, must be subordinated and kept in their due place; they are matters on which all thinking men will tend to differ, at least in details; it is upon these points that we split. There will always be the opportunist group, the uncompromising revolutionary group, and the purely educational groups in the movement. But in the Labour Church all these must meet and be fused, as it were, by the heat of common feeling for common ideals, into one force.[60]

After 1900, the Labour Church declined and as Unwin became increasingly involved in his architectural schemes his ethical work lapsed. His beliefs however remained intact. After the move to the fledgling Letchworth Garden City in 1904, where no church had yet been built, the large living-room in the Unwins' house was the setting for 'Sunday Evening Services' which, like the Labour Church services of the 1890s, were of a non-denominational nature 'such as could be joined in by those of any, or no, creed'[61]. The services were organised by Bruce Wallace, formerly of the Fellowship of the New Life, who had also been active in the Labour Church in the 1890s. Lecture notes survive for an address given to this group by Unwin around 1905; the subject, appropriately, was the ethical ideas of Hinton[62].

From the time of his first visit to Carpenter's Millthorpe Unwin retained his belief in what he called 'social experiments'. In 1887 in *Commonweal* he affirmed:

> Unlike some Socialists, I hope for great results … from individual efforts, from men who try to practise common interests and to do useful work, from small societies for working co-operatively in farming and other branches of industry – in short, from all who, by living, in spite of conditions, as far as possible in accordance with their ideal of life and society, are helping to spread the ideals and sentiments which will make our life in the future happier than it has been in the past.[63]

Between 1896 and 1904, Unwin pursued the architectural implications of this belief. At the level of the individual home, it shaped the Unwins' family life at Chapel-en-le-Frith and the recommendations on house design made in *The Art of Building a Home*. At the level of the model community, it shaped the ideas presented by Unwin in article form in 1900–01 and embodied in the design of Letchworth Garden City in 1903–04.

Unwin's thinking about the home centred on the simple life. In part the simple life was a development of the injunction issued by Ruskin in 'The Nature of Gothic' to purchase only such goods as were the product of undivided labour. It was this point that Unwin put in a lecture at Sheffield in 1897:

> As workers our first thought should be of the conditions of work, of its effect on us, of our joy in our work…. As consumers we should think of the lives of the men we consume. Everything we buy, every service we accept, is so much human life & it depends on us largely whether we buy & consume the joyous happy hours of labour or the weary hours of drudging toil.[64]

At their cottage at Chapel-en-le-Frith, described by their friend and neighbour Katherine Bruce Glasier, the Unwins carried this out enthusiastically, with curtains, blankets and clothes made from 'Ruskin' flannel, hand-woven in the

Isle of Man, and wrought-iron fittings 'that had brought the village blacksmith a new delight in life'[65].

But more importantly the simple life drew on the radical antithesis between convention and real needs established by Carpenter in the 1880s. In place of the conventional lifestyle of the wealthy, which in this view enslaved both those on whose labour it depended and those who were its supposed beneficiaries, Unwin called for 'a simpler form of life, one which need not cost so much of the labour of others to maintain, or so much of their own to produce'[66]. The application of this notion of the simple life to domestic architecture was the main theme of *The Art of Building a Home*. As one reviewer noted, the authors:

> insist on the importance of designing a house with a primary view to the comfort and convenience of those that will occupy it, and of making it as far as possible a truthful expression of the life that is to be lived there, instead of a mere echo of conventions.[67]

In *The Art of Building a Home* Parker & Unwin took as their target the ordinary house which, 'sacrificed to convention and custom, neither satisfies the real needs of its occupants nor expresses in any way their individuality'.[68] 'Real needs', according to Parker & Unwin, started with nature, and it was from nature, in the form of the site, that the design of the house should proceed: 'The site is the most important factor to be considered, for it usually suggests both the internal arrangement and the external treatment.'[69]

Nature's influence operated on the design in three ways: first, aspect (every living-room had to be open to the sun); second, prospect (the best possible view had to be secured from the main rooms); and third, landscape (the house had to adorn rather than deface the landscape in which it stood). In urban or suburban building this meant that merely conventional ideas, such as 'the convention that a cottage should face to the street'[70], would have to be abandoned, and the arrangement of roads and buildings, as much as the internal plan of the house, be suited to these dictates of nature. In language reminiscent of Ruskin's advice in *Modern Painters* to 'go to Nature', Parker & Unwin stated that:

> to produce a good plan, one should go to the site without any preconceived conventions, but with a quite open mind, prepared to receive … all the suggestions the site can offer.[71]

Beyond that it was a question of attending to the life and real needs, as opposed to the 'conventional wants', of those who were to live in the house. In the case of a labourer's cottage, this would produce a radical transformation in the internal plan, eliminating the parlour (demanded only by 'a false convention of respectability') and replacing it with a single large living-room, with the right aspect and a good view, and containing everything needed for daily living from the

Fig. 56 Parker & Unwin, design for Unwin cottage at Chapel-en-le-Frith, 1897, (right), interior perspective, and Fig. 57 (below), plans, from The Art of Building a Home, *1901*

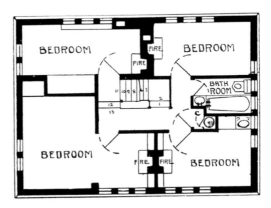

FIRST FLOOR PLAN

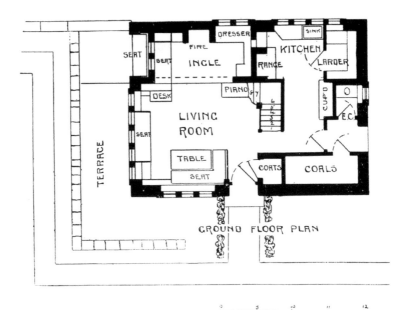

GROUND FLOOR PLAN

cooking-range to the piano[72]. A middle class house too would be closer to this
simple cottage than to a conventional villa:

> Those whose main desire is for beauty in their lives are coming to see that to the
> rational cottage as sketched above, with its ample living-room and the other
> necessaries of a decently comfortable life, they must add with great caution
> and reluctance, and only as dictated by really pressing needs.[73]

The result would be something closer to a Japanese interior as described by the
orientalist Lafcadio Hearn than to a conventional English home.

The Art of Building a Home included an unexecuted design made about 1897
for a new cottage for the Unwins on a site at Chapel-en-le-Frith (Figs 56 and 57).
As with other Parker & Unwin projects of this period, the design was a good
deal less assured than the prose and, for all the talk of simplification, for a small
house the form was one of considerable complexity. (When challenged at the
Society of Architects on the way in which their planning cut rooms up 'into
nooks and crannies', Unwin replied rather ungenerously that the 'sketches were
his partner's work, and most of the originality was due to him'[74].) Nonetheless in
Unwin's eyes the plan represented the expression of the simple life, with conventional ideas, such as the upstairs/downstairs distinction between parlour and
kitchen, eliminated and everything designed according to need and purpose, 'so
that there is no incongruity between the desk and the dresser, the piano and the
plate rack'[75].

Although to this extent Unwin's notion of the simple life as set out in *The
Art of Building a Home* followed Carpenter, one important difference should be
noted. For Carpenter, a crucial element in the simple life was the break with the
convention not just of manners and surroundings, but also of heterosexuality
and the family – a notion entirely absent from Unwin. Whereas for Carpenter
the simple life meant in part the opportunity to live more or less openly in a
homosexual relationship, for Unwin this notion of a transformation of gender
and family roles was absent. While proposing a transformation in the physical
setting of the home, Unwin's writings tended to reinforce rather than challenge
the conventionality of the social relations that were to exist within it.

Whatever might be achieved by individual attempts to lead a higher form
of life, more might be expected from social experiments in which not just an
individual or family but a whole community was involved. To Unwin socialist
colonies such as the Moorhays colony constituted the germ of the socialist life of
the future[76]. In 1889, describing a Sunday outing by the Chesterfield socialists to
the early eighteenth-century Sutton Hall, Unwin wrote:

> Small wonder that, as we stood looking at the house and the splendid view it
> commands, we should fall to talking of the 'days that are going to be', when
> this Hall and others like it will be the centre of a happy communal life. Plenty of

room in that large house for quite a small colony to live, each one having his own den upstairs ... and downstairs would be large common dining-halls, dancing-halls, smoking rooms – if indeed life shall still need the weed to make it perfect.

And we chatted on, each adding a bit to our picture; how some would till the land around and others tend the cattle, while others perhaps would start some industry, working in the outbuildings or building workshops in the park, and taking care not to spoil our view. Others, again, could work in the mines and bring up coal, of which there is a good supply just now being worked by a neighbouring company....[77]

The 1890s was a great period for communitarian experiments that sought to realise visions of this kind. One of the Unwins' closest friends, the ethical socialist Katherine Bruce Glasier, was involved in the early 1890s in one of the most famous, the Starnthwaite colony near Kendal. The late 1890s saw a new wave of communitarian experiments, in part triggered by the setback received by the Independent Labour Party in the 1895 election; the English Tolstoyans, led by JC Kenworthy and Bruce Wallace of the Fellowship of the New Life, were involved in several colonies (Purleigh, Whiteway and others) where they attempted, as one report put it in 1898, 'to live Socialism'. Nor was this a purely insular phenomenon: in Leeds in 1897 an outbreak of 'Cosme fever' was reported, with local socialists being attracted by the lure of the socialist colony in Paraguay[78]. By the turn of the century many of these ventures had collapsed from inadequate organisation, and the rump of the Tolstoyan group transferred their allegiance to what appeared to be the more promising venture of Ebenezer Howard and his Garden City Association, founded in 1899.

Letchworth Garden City, based on Howard's ideas, was a more highly organised version of the simple life colonies of the 1890s. Howard had approached Unwin for a design early in 1903 and (following a competition which Lethaby and Ricardo also entered) Unwin's plan for the new city was formally adopted a year later. Early settlers at the new venture along with the Unwins included veterans from the colonies of the 1890s, including Bruce Wallace and Carpenter's erstwhile partner George Adams, who brought with him the sandal-making business he had started at Millthorpe. According to Charles Lee, who moved to Letchworth in 1907, a 'typical Garden Citizen':

Wears far-and-near spectacles, knickerbockers and of course sandals.... Vegetarian and member of the Theosophical Society.... Over his fireplace – which is a hole in the wall lined with brick – is ... a large photo of Madame Blavatsky. Some charming old furniture, several Persian rugs etc. Books – works of Kipling, Lafcadio Hearn, 'Isis Unveiled', Wm Morris, Edward Carpenter, HG Wells, Tolstoi, etc.

To those unfamiliar with the socialist version of colonial life, Letchworth was quite confusing; 'a morris dance of many-coloured movements – Buddhism, Theosophy, Christian Science, Female Suffrage, several brands of Socialism and who knows how many varieties of Vegetarianism?'[79]

It was in the context of this tradition that Unwin developed the ideas about the design of socialist colonies that he presented in his paper 'Co-Operation in Building', published in article form in December 1900 and January 1901[80] and reprinted in *The Art of Building a Home*. In his earliest surviving lecture, 'The Dawn of a Happier Day', written under the influence of Hyndman and Carpenter in 1886, Unwin had derided the notion of a 'Golden Age' in the past, stating that there had been much violence, ignorance and drudgery in the Middle Ages and that socialists should look for their Golden Age to the future[81]. By 1900, however, he had come to share the medieval enthusiasm of Morris and regarded the Middle Ages as the model for the architecture of communitarian experiments. What made the buildings of the middle ages – the village, the manor house, the college quadrangle – so appealing, according to Unwin, was the orderly social life that they expressed:

> The village was the expression of a small corporate life in which all the different units were personally in touch with each other, conscious of and frankly accepting their relations, and on the whole content with them. This relationship reveals itself in the feeling of order which the view induces....[82]

The socialist colonies would revive the collective life of the Middle Ages which had been shattered in the period of individualism:

> association for mutual help in various ways is undoubtedly the growing influence which is destined to bring to communities that crystalline structure which was so marked a feature of feudal society, and the lack of which is so characteristic of our own.[83]

Accordingly for their architecture the colonists should revive the building forms of the Middle Ages: 'Why should not cottages be grouped into quadrangles, having all the available land in a square in the centre?'[84] Or, if a full quadrangle did not fit the site, houses could be built in just three sides of the quadrangle, an arrangement that Unwin had found in the manor houses of the late Middle Ages[85].

In these forms a new life would develop. Relationships between people would change: as well as domestic life in the individual cottages, each quadrangle would provide for communal life by including common rooms for relaxation, baking, laundry and bathing:

> From this to the preparation of meals and the serving of them in the Common Room would be only a matter of time; for the advantage of it is obvious. Instead

of thirty or forty housewives preparing thirty or forty little scrap dinners, heating a like number of ovens, boiling thrice the number of pans and cleaning them all up again, two or three of them retained as cooks by the little settlement would do the whole, and could give better and cheaper meals into the bargain.[86]

The relationship to nature would change also: instead of the land being divided up into individual private plots of a diminutive area, it would be grouped into the central space of the quadrangle where it would constitute a large garden for collective use. Thereby:

association in the enjoyment of open spaces or large gardens will replace the exclusiveness of the individual possession of backyards or petty garden-plots....[87]

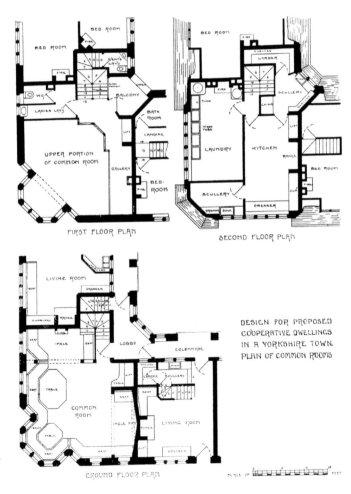

Fig. 58. Parker & Unwin, design for a quadrangle of co-operative dwellings, 1898-99, from The Art of Building a Home, 1901

DESIGN FOR PROPOSED COOPERATIVE DWELLINGS IN A YORKSHIRE TOWN. PLAN OF COMMON ROOMS

Fig. 59. Parker & Unwin, design for a quadrangle of co-operative dwellings, 1898-99, part-plan showing double-height common room and other communal facilities, from The Art of Building a Home, *1901*

To illustrate these ideas, Unwin published two possible arrangements for co-operative dwellings. One (designed for a site in Bradford in 1898–99) was of the quadrangular form that Unwin proposed for general use, with individual dwellings on four sides and common rooms occupying one corner, and the central open space laid out as a garden or lawn (Figs 58 and 59). Two versions of this plan were shown, one with three-bedroom cottages and the other with five-bedroom houses. The other scheme illustrated by Unwin was a design for a south-facing slope: a group of houses arranged informally around three sides of a green. This design originated in a tentative commission received in 1899 from Isabella Ford of Adel Grange near Leeds, a friend and disciple of Carpenter's.

Unwin approached the design of Letchworth Garden City in 1903–4 as essentially an elaborated version of a co-operative scheme. He discussed the housing for the garden city in the paper he gave to the Bournville conference of the Garden City Association in September 1901: the design was to follow the ideas set out in *The Art of Building a Home*, that is, being based on the 'real needs' of nature and social life rather than mere convention. The planning of the city followed the same notions, paying attention both to nature (not just aspect and prospect, but also topography and the preservation of existing trees), and to the 'real needs' of the occupants (such as access, employment, economy). Ownership of the land by the community – one of the unchanging fundamental tenets of Unwin's political philosophy since the 1880s – would mean that at the garden city the community for the first time 'secured freedom to express its life adequately' in its architecture[88].

Art and the city

In his paper to the 1901 Garden City Association conference, Unwin called for the municipal authority in the garden city to lead the way in building housing. By this date there was a widespread demand for municipal provision of housing from trades councils and other bodies in the labour movement, including the Fabian Society, the Workmen's National Housing Council (set up in 1898) and above all the Independent Labour Party (ILP). The ILP, launched in Bradford in 1893, saw in housing a strong argument for its policy of labour representation on local councils and in parliament. The first issue of the Leeds ILP's *Labour Chronicle* in May 1893 carried an article on 'Unhealthy Dwellings':

> The disgraceful action of the Corporation in regard to back-to-back houses should prove a wholesome lesson to the working classes of Leeds; and we hope it will open their eyes to the blind folly of sending the monied classes to represent them.[89]

This journal also carried a front page article on 'Labour May Day' by Unwin. Unwin was sympathetic to the ILP from the outset but was doubtful how much a socialist political party could really achieve. In 1900, however, the ILP received a crucial boost when it gained the adherence of the trade union movement and shortly afterwards Unwin declared for the new party. In January 1902 an article in the *ILP News* (which was edited by his friend Bruce Glasier) set out Unwin's position.

Writing as a former 'member of the old Socialist League', Unwin declared that times had changed since the days of the League. The League's belief in the coming revolution had been proved wrong: 'the change was not destined to come in the way we thought and feared'. Now the idea of socialism was widely accepted and people worked to advance it in many different ways, not least through 'municipal and political life'. Unwin still believed that the political route held 'immense dangers ... dangers of corruption and compromise': it was essential, therefore, that 'if we are to enter this life we must follow a policy that has at least some chance of success'. The alliance with the trade unions offered 'the best chance we ever had of getting a Socialist Labour Party in Parliament' and, if the parliamentary route was to be followed at all, this alliance should be pursued determinedly[90].

Within the arguments of the time Unwin's article was an attack on the Social Democratic Federation and its notion of a 'pure' Socialist political party and a defence of the ILP's attempt to form a Labour Party in which socialists and trade unions would co-operate – the strategy that Mahon had advocated to Unwin's Clay Cross socialists in 1887 and the one on which the Labour Party was to be founded. In an editorial Bruce Glasier commented of Unwin's article:

> Although idealist as ever in his conceptions of the goal of Socialist effort, and although by disposition recoiling from the turmoil of electioneering, he frankly conceded the main ILP contention that the municipal and political Labour way is the right road to practical Socialist achievement.[91]

Keir Hardie, one of the main instigators of the alliance with the trade unions, was sufficiently impressed by Unwin's article to poach it for his paper, the *Labour Leader*[92].

Unwin's ILP article came shortly after he had presented his ideas on municipal housing at three major gatherings: in a speech to the Garden City Association conference (20 September 1901); in a lecture to the Workmen's National Housing Council (4 November 1901); and in a lecture to the Fabian Society (22 November 1901). This last, entitled 'Light and Air and the Housing Question', was subsequently published as Fabian Tract 109, *Cottage Plans and Common Sense*. In these presentations Unwin argued not just that municipalities should provide housing, but that the housing they provided should be of a new and quite different sort, based on the Unwin-Carpenter notion of a rational cottage answering to the real needs of the simple life.

In keeping with his argument that pragmatic policies could be justified only by success, Unwin approached the question of municipal housing in a resolutely practical way. Most local authorities acting to relieve the housing shortage, he said, would have to build neither in town nor country, but in the 'great suburban districts' in between, 'where, after all, the majority of working folk are housed'[93]. This meant that densities could scarcely be reduced below 24 to 36 houses per acre [0.4 hectare]. Since under the system of local authority finance, the capital cost of construction would be repaid over a long period, anything that was built should be the 'very best that is known', in other words 'based upon the permanent and essential conditions of life and health, not on passing fashions or conventions established by the speculative builder.'[94] This argument, linking Carpenter's distinction between convention and real needs to the demands of local authority finance, was later to figure prominently in Unwin's contribution to the post-war debate on standards for state housing.

The design of the municipal house, Unwin wrote, 'had to be thought out from the beginning, as though no custom in connection with such buildings had ever grown up'[95]. The needs of the occupants were of three sorts. First, their needs with respect to nature: essentially sun, light and air. This meant abandoning 'the convention that a cottage should face to the street'[96] and arranging the layout of roads and houses according to aspect and prospect, eliminating backyards and back-projections in the process. Second, their needs as members of a family. This meant ensuring a large and workable living-room, a scullery, larder, coal-store, toilet, and bedrooms, but not a parlour (not 'necessary to health or family life') nor a separate chamber for the entrance and stairs ('an extreme instance of valuable room and air space sacrificed to thoughtless custom and foolish pride')[97]. Third, there were the needs of the occupants as members of a community. These were to be met by the provision of communal centres in the corner of each quadrangle, which would provide those facilities that were otherwise available only to the wealthy (laundries, reading-rooms and if necessary baths) and would foster the growth of the 'co-operative spirit', as befitted a municipal undertaking (Fig. 60)[98].

The publication of Unwin's lecture as a Fabian Tract was seen by the Fabian Society (which Unwin had joined only in March 1901)[99] as 'a new departure, both in respect of the introduction of plans and sketches ... and in the subject matter,

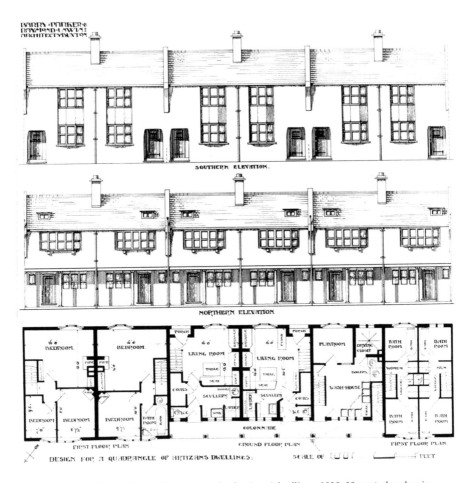

Fig. 60. Parker & Unwin, design for a quadrangle of artisans' dwellings, 1898–99, part-plan showing dwellings and communal wash-house, bathrooms and playroom, from Cottage Plans and Common Sense, 1902

which is only indirectly political'. Nonetheless the Society was pleased to report that the tract 'has been very widely noticed in the Press, and a very unusual number of orders for single copies have been received'[100]. In retrospect the interest provoked by *Cottage Plans and Common Sense* does not seem surprising, for it marked a turning-point in architectural thinking about housing design.

In the final sentence of *Cottage Plans and Common Sense* Unwin spoke of the 'simple dignity and beauty in the cottage' that was necessary to 'the proper growth of the gentler and finer instincts of men'. Throughout his writings of the period 1901–04, reference was made to the indispensability of art in housing and urban layout. Nowhere was the 'claim of beauty or art' more important, Parker & Unwin wrote in 1903, than in the 'transforming of open fields into dwelling places'[101]. In 1906 in a lecture on city planning at Cambridge, Unwin argued that:

the introduction of civic art, of the due consideration of the element of beauty, was the most important and most urgently required reform in town affairs at the present time.[102]

This emphasis on the role of art in the building of houses and towns was explicit in the title of *The Art of Building a Home*. It was also evident in the original title proposed for the book that Unwin wrote to coincide with the town planning legislation of 1909: until shortly before publication this was to have been called, not *Town Planning in Practice* [CD] but *The Art of Town Planning*[103].

What did Unwin understand by 'art' in this context? A number of ideas contributed to his conception of art, but the most important in regard to the city was the Ruskinian idea of art as expression and particularly Carpenter's transcendentalist notion of art as the expression of life. One of the quotations from Carpenter cherished by Unwin was the following:

> Art is expression: expression of that which is else inexpressible. In all true art, whenever we see beauty, something passes to us, some touch of that which is infinite: something from a kindred soul to ours.[104]

This notion was impressed on Unwin by Carpenter at an early stage of their relationship:

> One early difference I recall, as the incident has significance.

> When he was just in Millthorpe, we coming down the hill from Freebirch, I felt that the slate roof did not quite belong in that lovely valley.

> He felt I had not understood his aim. Essentially seeking expression: not decoration. In building Millthorpe E.C. sought expression of life....[105]

In dealing both with the house and the city Unwin employed this notion of art as the expression of the life of the inhabitants. In regard to the house, we have seen that in *The Art of Building a Home* Parker & Unwin insisted that expression be given to the individuality of the occupants. Likewise Unwin regarded the city as a work of art in which the collective life was expressed:

> fine city building is an art. The city is a form in which the life of its people expresses itself.

It was therefore subject to the same rules as any other art:

> In this art, as in any other, success depends on there being something fine to express and upon it being finely rendered.[106]

The role of the designer or town planner was to act as the medium in this process, like a brush in the making of a painting. Unwin's notes for a lecture in 1908 on 'town planning' (the term came into general use in 1906) made the point succinctly:

> Demand for Town Planning [the] result of common life seeking expression. Essentially a Civic Art, designer being channel through which it expresses itself – the brush with which they paint.[107]

For the characterisation of the art that the designer was to realise, Unwin drew on Lethaby and the notion of art as the 'well-doing of what needs doing'. The common life of the community would find expression not at the expense of the practical needs of the people, but by meeting those needs with that 'small margin of generosity and imaginative treatment that constitutes it well done'[108]. It was this that gave city planning its value:

> It is just the little margin of imaginative treatment which transforms our work from the building of clean stables for animals into the building of homes for human beings, which is of value; for it is just this which appeals to and influences the inner heart of man.[109]

The city could be reclaimed for art in this way only if the community took for itself the necessary powers over the control and ownership of land. In the Middle Ages, Unwin believed, the land had belonged to the people: the history of the land since then was simply 'a history of the confiscation of the people's rights'[110]. These rights had to be restored, through the collective ownership of land, if the community was to be in a position to determine the form taken by the town. In the garden city land was owned collectively; similar 'powers to purchase and hold land around the town should also be obtained for our municipalities'[111]. For only by the municipal control of land could the claims of beauty be made to tell:

> So long as we leave individual landowners to develop their own plots of land in their own way, our towns must continue to grow in their present haphazard manner. But if their development is arranged and controlled by some central authority, it becomes at once easy to consider the possibilities of the site, to preserve features of beauty and interest, to keep open distant views, and to arrange roads with proper regard for convenience and beauty.[112]

Thus, while proposing a role for the state which Ruskin would have found hard to accept, Unwin's work in housing and city planning conformed fully to the goals of art as defined by Ruskin. Unwin was extending the borders of the category of art as defined by Ruskin to include the city, giving the city a place alongside the painting, the work of sculpture and the work of architecture as something capable of expressing, and speaking to, the human spirit.

Unwin and 'The Nature of Gothic'

In the end, then, Unwin belongs to the Ruskinian tradition, but only indirectly and at a distance. Between Unwin and the tradition based on 'The Nature of Gothic' lay the cultural and political distance that separated the north from Oxford and London; between Unwin and Ruskin stood Carpenter. In the period of Unwin's intellectual formation, Ruskin was a significant but remote influence. To a socialist activist of the 1880s, Ruskin was an ambiguous, if not downright reactionary, figure, with what Unwin called his 'onslaught onto Democracy, especially in Fors Clavigera'; in 1885 the prospect of addressing a group of Ruskinites, 'who are Christian socialists but opposed to organisation'[113], made Unwin decidedly nervous. Moreover, while generally sympathetic to Ruskin's social and ethical teaching, Unwin had too much of Carpenter's positive attitude to the labour-saving potential of machinery to agree with Ruskin's proscription of the machine: the Ruskinian ideal, he wrote in 1885, was 'a very good one in the main though as to his theories of machinery probably we cannot send the clock of time back again....'[114]

It was in Morris that Unwin found the most persuasive presentation of Ruskinian ideas. Unwin recalled:

> Underlying and partly promoting Ruskin's love of Gothic architecture, this point of view emerged more clearly in the words of William Morris. It became indeed one of the moving impulses of his life. To Morris, art was the expression of man's joy in his work; and his life was spent in exploring the endless possibilities of such enjoyment, and was completed in a desperate attempt to secure for all men the kind of work in which some gladness may be found, and the conditions of labour in which it may be enjoyed.[115]

Under Morris's influence Unwin for a time became an advocate of the view that free and joyful labour should form the basis of socialism, arguing the point at length in a lecture given at Sheffield in 1897, 'Gladdening v. Shortening the Hours of Labour'[116]. Within the pages of *The Art of Building a Home* Unwin reiterated the lesson of Gothic architecture as originally drawn by Ruskin:

> In fact, we read in these old buildings, as in an open book, of a simple workman who was something of an artist, one who could take pleasure in his work, finding joy in the perfection of what he created, and delight in its comeliness.
>
> Whenever we again raise up such an army of builders, working at their trades with the pleasure of artists, then will all buildings become as beautiful as ... our old cathedrals and abbeys.[117]

And throughout his life he continued periodically to pay obeisance to this view. At the Lethaby memorial evening at the RIBA in 1932, Unwin referred to the belief underlying 'the joy which Lethaby and William Morris took in Gothic art: that is, their belief that it gave great opportunities for enjoyment to the workman' and affirmed that 'I still retain the conviction that some day we shall again find a style of building which will afford an opportunity for joy to all the workmen who are engaged on it', although he admitted that 'we do not seem to be approaching much nearer at the present time'[118].

In practice, however, the notion of 'joy in labour' was peripheral to Unwin's thought. The most telling effect that Morris had on Unwin's architectural thought was his medievalism: under Morris's influence Unwin abandoned the antipathy to the Middle Ages evident in his 1886 lecture on the theory of surplus value, and in *The Art of Building a Home* and the other writings of the early 1900s argued for the Middle Ages as the architectural model for socially progressive building. Morris may also have been responsible for leading Unwin away from his initial interest in the 'scientific basis' to socialism given by Marx and represented in English socialist politics by Hyndman. In 1902 Unwin, recalling the 'revolutionary' days of the mid-1880s, depicted Hyndman and Morris as alternative faces of socialism, offering respectively the materialist and the ethical view. There was Hyndman, with 'his top hat, frock coat and general air of respectability':

> We admired his ability, we respected his pioneer work, we felt that the Marxian theories were great, though we did not presume to understand them…. Then, too, how we loved our Morris when he came to us sharing our illusions, full of life and joy, caring as little for the value theory as we did, but very much in earnest….[119]

In general, however, Morris's effective influence on Unwin was limited by the political and geographical distance that separated them. Although his Oxford home and school days gave Unwin a bridge with the southern tradition, his intellectual formation took place in the north, where industry rather than crafts prevailed and – particularly among the factory districts of West Yorkshire – the labour movement was strong. Here the working class looked to self-organisation in trade unions and to the potential of the state in order to gain better conditions of life and work, and had little time for the policy of abstention that appealed so strongly in the south. The northern socialists also showed a longstanding interest in socialist colonies as a way of delivering immediately some of the benefits of socialism. The result was that by the early 1900s Unwin looked to colonies such as Letchworth and even more to municipal initiatives in housing and town planning as the basis for his architectural prescriptions, accepting what Bruce Glasier termed 'the municipal and political Labour way' as the best means of socialist advance.

In the industrial north, the belief in labour as the major source of human satisfaction had less appeal than in the non-industrial south. Rather, as Unwin put it in 1888, what people wanted was 'to produce enough wealth to keep us in comfort with as little labour as possible'[120] – something much closer to the Fabian view. Hence, for the most part Unwin followed Carpenter in seeing socialism in terms of the reduction of toil, enabling 'each man to produce in the easiest way known'[121], and he looked forward to 'the increased leisure which only Socialism can make possible'[122]. This attitude to pleasure and labour was fundamentally at odds with that of the Ruskinian tradition based on 'The Nature of Gothic'.

Chapter eight

Unwin and Sitte

In 1904 Raymond Unwin read Camillo Sitte's *City Planning according to Artistic Principles* – not the German 1889 original but the (bowdlerised) 1902 French edition by Camille Martin[1]. What was Unwin's reading of Sitte? Unwin's major theoretical text, *Town Planning in Practice* (1909) [CD], reveals a contradictory attitude. On the one hand Unwin was enthusiastic in his adoption of Sitte's notions of enclosure and the street picture (in Sitte, *stadtbild*; in Martin, *tableau*). On the other hand, Unwin criticised Sitte repeatedly and at length on the question of informality and irregularity in design.

As George and Christiane Collins have shown, the Martin edition was very different from Sitte's original, incorporating not only different images but also an entirely new (and highly influential) chapter on streets. Whereas for Sitte the street had no artistic merit (and was therefore unworthy of attention), in the Martin edition the street became a central concern; in the Collins' judgement, therefore, the Martin edition was 'a completely different book, not only poorly translated, but actually enunciating ideas that are dramatically opposed to Sitte's principles'[2]. But all this was unknown to Unwin. So far as he was aware, he was at the same time accepting, and rejecting, important elements of Sitte's thought.

Unwin's reading of Sitte was part of the internationalisation of urban and town planning debate of the period 1900–14, which has been well described by Anthony Sutcliffe[3]. But to understand it properly we must also locate it in the political economy of the time, centring on the garden city movement, and specifically within the ideas about the relationship between the form and social life of the city that Unwin had developed over the previous quarter-century. For it was the garden city movement's concern with low-cost housing, and Edward Carpenter's ethical teachings, that shaped Unwin's reading of Sitte.

Political convergence and the garden city movement
In the second half of the nineteenth century Britain lost its global economic hegemony and started to fall behind its rivals, particularly Germany. This rivalry underlay the imperial adventurism of the 1880s and 1890s (including the South African War of 1899–1902) and was to lead to British naval rearmament from 1903 (intensified by the huge Dreadnought programme launched in 1909) and eventually to war in 1914. Although Anglo-German rivalry was essentially a conflict between two free-market economies, it was recognised in Britain from as early

as the 1880s that significant modifications to the free market – state provision as opposed to state regulation – were needed if terminal national decline was to be averted. The discovery in 1899 that three-quarters of would-be recruits from Manchester, one of Britain's largest cities, were physically unfit for military service created shock waves that reverberated for several years[4]. Particularly alarming in this connection was the discovery that Germany was far ahead of Britain in terms of action by the state to promote the physical well-being of the population. TC Horsfall's *The Improvement of the Dwellings and Surroundings of the People: The Example of Germany* (1904) presented the evidence and the conclusion to startling effect. As Horsfall later put it, unless:

> we protect the health of our people by making the towns in which most of them now live, more wholesome for body and mind, we may as well hand over our trade, our colonies, our whole influence in the world to Germany, without undergoing all the trouble of a struggle in which we condemn ourselves beforehand to certain failure.[5]

The lead in calling for a shift from free-market liberalism to some form of welfare provision was taken by industrialists from the new mass-consumption industries. As manufacturers of household goods and consumables (soap, chocolate and soft drinks), Lever, Cadbury, Rowntree and Idris knew that their success depended not on the poverty but on the affluence of the working population, and their involvement at Port Sunlight (1888–), Bournville (1895–), New Earswick (1902–) and Letchworth Garden City (1903–) sprang from a wider philosophy of welfare provision. At first the state stayed on the sidelines, but from 1907 it began to take an interest in assisting these initiatives to improve 'the housing and surroundings of the people' and before 1914 had commissioned two reports on low-cost housing involving the leading designer from the garden city movement, Raymond Unwin [CD][6]. With the onset of war, the state's interest in such initiatives accelerated markedly (see chapter one).

The shift by industrial capital towards welfare provision for the working class was mirrored by a complementary shift by the socialist movement, from the revolutionary socialism of the 1880s (the Social Democratic Federation and the Socialist League) to the 'practical socialism' of the 1900s (the Independent Labour Party and the Labour Party). Rather than agitate for a new society based on overthrowing the existing system, socialists of the 1900s looked to achieve tangible gains for the working class within the existing framework of society. As Raymond Unwin, himself a former Socialist League activist, wrote in 1902 in the *Labour Leader*, the old belief in the imminence of revolution had been proved wrong: 'the change was not destined to come in the way we thought and feared'; instead 'as an idea people now accept' socialism, 'and quietly pass on to do something in one direction or another to translate the idea into life.'[7]

The garden city movement was one of these 'directions' of the 1900s, bringing together welfare capitalism and reformist socialism in a common goal. The great theme of the garden city movement was the new spirit of community and coop-eration. 'Our towns and our suburbs express by their ugliness the passion for individual gain which so largely dominates their creation', wrote Unwin. The task of the town planner was to 'do something to restrain the devastating tendency of personal interests and to satisfy in a straightforward and orderly manner the requirements of the community.'[8] Instead of replicating the unsavoury features of existing towns (squalor, ill-health, overcrowding, lack of open space and amenity), residential districts of a new sort would provide attractive housing for the working population and people of small means. By necessity these would be located where land was cheap, either on the edge of existing towns (garden suburbs, town planning areas) or in the country (the garden city).

Unwin's thinking about city planning proceeded from his encounter with Edward Carpenter and particularly Carpenter's concept of architecture as the expression of life (see chapter seven). The layout of residential districts should be 'based upon permanent and essential conditions of life and health, not on passing fashions or conventions established by the speculative builder'[9]. Instead of following the standard byelaw format, the layout should be determined by the site and the likely needs of the inhabitants (Fig. 32). To meet these, the town planner would need the right techniques (particularly the techniques of low-density layout developed by Unwin) but should not seek to impose *a priori* preferences about form. Against all advice in favour of either formality or infor-mality in design, Unwin insisted that the planner should 'avoid dogmatising on the theories and instead keep very closely in touch with actual requirements'[10]. The task for the planner was 'to find expression, not for some preconceived ideas of his own, but for the needs and life of a rising community'[11].

From *stadtbild* to street picture

It was in the context of these ideas that in 1904 Unwin read the French edition of Sitte. He immediately recognised the importance of 'Camillo Sitte's great book', not just in terms of its influence on German planning but in its own right[12]; in *Town Planning in Practice* he regretted that the plan of Letchworth Garden City had been executed before he 'had the good fortune to come across Camillo Sitte's book'[13]. Those parts of Sitte that were irrelevant to his purpose Unwin simply passed over – most notably Sitte's contention that the ordinary districts of towns were 'artistically unimportant' and could be ignored[14]. Two main portions of the book caught his attention: the concepts of enclosed urban space and the street picture; and what he saw as the argument for irregularity and informality in the planning of streets.

For the latter, notwithstanding Sitte's reputation, Unwin had no sympathy. One of the major themes of *Town Planning in Practice* was Unwin's attack on the fashion for what he saw as excessively picturesque planning: as he put it, 'whatever the character of the street, it is of the utmost importance to avoid mere aimless wiggles'[15]. An *a priori* preference for curving streets and irregular lines was in Unwin's view no different from any other sort of 'convention' that was not based on 'real needs'. According to Unwin, it was 'maintained by Sitte and others that much of the irregularity characteristic of the medieval town which we have been apt to consider the result of natural and unconscious growth was, on the contrary, deliberately planned'[16]. Unwin disagreed with this interpretation; it was much more probable, he said, that these irregularities were an instinctive response to site and other conditions:

> The difference between this instinct which made the best use of irregularities, and the conscious artistic designing of those irregularities, may seem a small one, but it is of importance when upon it is based the claim that the conscious designing of the modern town planner should be carried out on the same irregular lines.[17]

This claim (which the Collins have shown was more the work of the 1890s followers of Sitte than of Sitte himself) Unwin categorically rejected: irregularities in streets and building lines should be used only where there was 'definite justification'[18].

In contrast, Sitte's notions of the enclosed plaza or *place* and the street picture were adopted by Unwin with enthusiasm. 'It was not until Camillo Sitte drew attention to the artistic side of town planning in his book *Der Stadtebau* that the true meaning and importance of the *place* was realised,' he wrote; 'Camillo Sitte devoted a large part of his volume to the examination of *places*, and to elucidating the principles of their design.'[19] Unwin then proceeded to devote a substantial portion of his volume to summarising what Sitte had written, complete with illustrations drawn from the Martin edition (Fig. 61) and elsewhere.

Fig. 61. Raymond Unwin, plan and perspective of the Rue des Pierres, Bruges, based on the Camille Martin edition of Camillo Sitte, from Town Planning in Practice, *1909*

Nearly 40 pages of *Town Planning in Practice* were given over to summarising Sitte on centres and enclosed *places*: 'A *place* ... in the sense in which we wish to use the word should be an enclosed space. The sense of enclosure is essential to the idea.'[20] In ancient towns entrances to the *place* were 'so arranged that they break the frame of the buildings very little if at all', with the corner entrances to the Piazza Erbe in Verona disposed 'in such a way that when looking across the *place* no direct view down either street is open'[21] (Fig. 62). At the Marienplatz in Munich enclosure was secured by 'diverting the course of the road immediately after its leaving the *place*'[22]. Public buildings were located not in the centre but at the side, often attached to surrounding buildings. 'Camillo Sitte quotes the almost universal custom of the ancients to prove that buildings are not seen to the best advantage when seen in isolation'[23]: 'this is not the way to produce satisfactory pictures or to show the buildings to the best advantage'[24]. In shape and proportion, the *place* should relate to its principal building: as Sitte said, deep when the building was tall (eg a church) and wide when the building was broad (eg a town hall). As well as these major central *places*, there were also small, simple *places* that could be created at road intersections 'by breaking the line of direction, the result being that the view down each street is closed and a figure resembling a turbine is produced'[25] – a point made with illustrations 'taken from Camillo Sitte's book'[26] (Fig. 63).

The discussion of streets involved Unwin in the critique of Sitte already mentioned. This did not prevent him, however, from discussing the street in visual and spatial terms derived from Sitte. 'We have seen in speaking of *places*

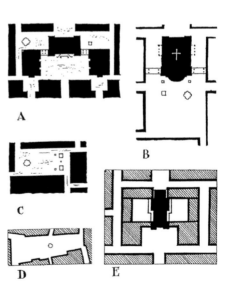

Fig. 62. *Piazza Erbe, Verona: plan from Camillo Sitte reproduced in* Town Planning in Practice, *1909*

Fig. 63. *Raymond Unwin, 'Places and groups of places adapted to modern conditions, as recommended by Camillo Sitte', from* Town Planning in Practice, *1909*

and squares how important to the effect is a sense of enclosure, the completion of the frame of the building; and much the same applies to street pictures'[27]. This meant thinking about the street as a series of views, to be manipulated and controlled by changes and shifts in the street lines. To 'secure a fairly frequent completion of the street picture, we shall desire to close from time to time the vista along the street; this result is secured by a break in the line of the street; or by a change of direction, or curve ...'[28] (Fig. 64).

Fig. 64. Raymond Unwin, 'Plan and sketch of street showing on one side the uninteresting vanishing perspective of the unbroken building line, and on the other the more picturesque result of breaks', from Town Planning in Practice, 1909

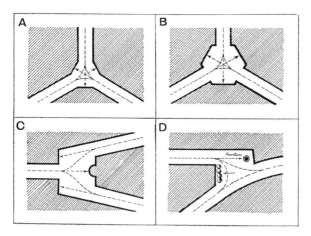

Fig. 65. Raymond Unwin, ways of creating an enclosed street picture at street junctions, from Town Planning in Practice, 1909

Road intersections offered a particularly important field for street pictures; as Unwin observed, it was 'upon the treatment of street junctions that much of the effect of the town will depend'[29]. This was one of the most original and important sections of Unwin's book, applying to the design of the 'broad mass of living quarters' the principles that Sitte had elucidated for the 'Sunday best'[30]. As he noted, 'once attention is given to the subject, there are very many ways in which street junctions can be treated' to produce satisfactory street pictures[31] and he went on to illustrate them (Figs 65 and 66), presenting some of the tech-

Fig. 66. Charles Wade, sketch of road junction (similar to Figure 65D), from Raymond Unwin, Town Planning in Practice, *1909*

niques for handling road junctions while preserving the street picture that he had developed at Hampstead Garden Suburb.

In his discussion of *places* Unwin turned from his summary of Sitte to reproduce an analysis of the centre of Buttstedt taken from the December 1908 issue of the periodical *Der Stadtebau*, with a series of street pictures drawn by his assistant and illustrator Charles Wade. The layout of this small town, with its enclosed space and its street pictures carefully maintained by shifts in the line of the street, offered important lessons for the modern planner (Fig. 67). Whether designed deliberately or not:

> we must admit the beauty of the effects produced and the success of the whole. Here we have a little town consisting of the simplest and plainest buildings in the main, and yet, owing to the splendid placing of its two public buildings and to the arrangement of its streets and places, the whole presents a degree of beauty and impressiveness quite astonishing ...[32]

This was the basis of Sitte's appeal for Unwin. Designing towns in terms of street pictures offered a way of creating beauty out of all proportion to the expense involved and was therefore appropriate to the social and economic objectives of the garden city movement. For the realisation of street pictures required not lavish expenditure on buildings, but the subsumption of the individual plot and the individual building under the control of the town planner as the representative

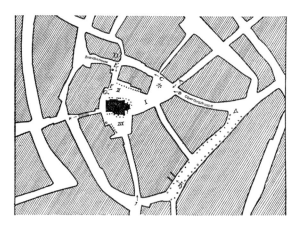

Fig. 67. Raymond Unwin/Charles Wade, plan (left) and street pictures (below) of Buttstedt, from Town Planning in Practice, *1909*

of the community. Only if the communal was recognised as taking priority over the individual could a layout designed in terms of street pictures be realised. The concept of the street picture that he derived from Sitte thus provided Unwin with the means he had sought for giving expression to the 'needs and life' of the community and, as such, for turning the city into a work of art.

Chapter nine

Rammed earth revival

If you go to the town of Amesbury in Wiltshire you will find, close to the former railway station, a settlement originally consisting of 32 houses, a quarter of which were constructed in earth materials. Various kinds of earth construction were used but the main type was rammed earth or *pisé de terre* (Fig. 68). Built in 1919–1921, the Amesbury development is the most tangible product of the adoption of rammed earth by the Board of Agriculture for the programme to settle soldiers and sailors on the land after the First World War. As well as Amesbury, the rammed earth revival of the early twentieth century gave rise to a number of publications. Perhaps most notable was the book on earth construction published in 1919 by Clough Williams-Ellis (who for a time was a salaried architect with the Board of Agriculture), *Cottage Building in Cob, Pisé, Chalk and Clay: A Renaissance* but there were also two government reports published by the Department of Scientific and Industrial Research, one (1921) by WR Jaggard specifically on Amesbury and the other (1922) by HO Weller on earth construction more generally [CD][1].

Eighty years later, as part of the worldwide interest in low-energy technologies generated by the eco-crisis, earth construction is much in vogue. The electronic catalogue at the RIBA library lists more than 200 publications from the past 25 years. Regular international conferences are held on the subject, including New Mexico (1990), Lisbon (1993) and Torquay (2000)[2]. Recently, rammed earth construction has been used by architects for prominent projects such as

Fig. 68. Ministry of Agriculture,
chalk-pisé cottage at Amesbury, 1920

the Chapel of Reconciliation built by Martin Rauch on the site of the former Berlin Wall and the visitor centre by Nicholas Grimshaw & Partners at the Eden Project in Cornwall[3]. The UK government has even returned to the subject, with a 30-month Department of Trade and Industry (DTI)-funded research project led by Peter Walker at the University of Bath started in 2002, which looked at the viability of rammed earth construction for social housing. The major outcome of this project was the publication in 2005 of a new technical guide to building in rammed earth, *Rammed Earth: Design and Construction Guidelines*[4].

For these present-day earth revivalists in the UK, the period around the First World War still forms the benchmark. Rowland Keable – a leading figure in the current rammed earth revival in the UK and the rammed earth contractor for the Eden Project and other UK projects – has stated that it was the Williams-Ellis book that opened his eyes to earth construction[5]. The editors of the *Terra 2000* volumes tell us that these publications dating from the early part of the twentieth century constitute Britain's main contribution to the large-scale investigation of earth construction[6]. Little wonder then that, 80 years after its first publication, the Williams-Ellis book has again been reprinted[7].

In today's climate – literal and metaphorical – it is largely assumed that the use of rammed earth like eco-friendly construction methods in general, is politically progressive and environmentally responsible. It is easy, therefore, to assume that this meaning inherently attaches to this form of construction. But what led the British revivalists of the early twentieth century to espouse this form of construction? Did earth construction then have the meanings and associations that it has today?

What is meant by rammed earth construction or *pisé de terre*? Essentially *pisé* is an exotic name for a form of wall construction that has been used for many hundreds of years in various parts of the world, including Africa, Asia and Europe. It differs from brick in that the earth is not baked at a high temperature in a kiln but is used raw. It differs from mud construction in that the material is used in a more or less dry form, rather than wet. It gains its strength not (as with mud construction) from being baked in the sun or being reinforced by a binding agent but from being compacted – ie rammed – using formwork similar to that used for in situ concrete. The usual manner of constructing walls involves making and erecting the formwork (of timber or steel) and laying the earth in courses around one foot [30.4 centimetres] deep. The earth is compacted by being rammed (traditionally by hand but today by machine) and the course is then allowed to dry for around three hours before the next course is laid. The earth wall has to be protected from water ingress from above and below by an overhanging roof and a base wall (of brick or concrete) about one foot [40.2 centimetres] high. Although the density of the wall offers protection against surface rainwater, in temperate climates such as the UK a 'raincoat' of some form such as lime roughcast or tar is usually considered necessary for habitable buildings[8]. The thickness of the wall required for stability (at Amesbury, 18 inches [45.7 centimetres] for the lower floor and 14 inches [35.5

centimetres] for the upper) and the thermal mass of the material leads to inherently good thermal performance, keeping the interior warm in winter and cool in summer.

Rammed earth was not an indigenous method of building in Britain, where the only indigenous form of raw earth construction was mud, reinforced with straw or some other binding agent. This was a traditional method used in Devon (and to some extent in south Wales), where it is known as cob; and in Norfolk, where it is known as clay lump. At the end of the eighteenth century, however, the *pisé* technique was introduced to Britain. The pioneering textbook on *pisé de terre* was published in Paris in 1790 by François Cointeraux and seven years later was translated into English by Henry Holland; thereafter the technology was included in nineteenth-century manuals in Britain, as elsewhere in the world. Unknown to Williams-Ellis and the other rammed earth revivalists of the early twentieth century, rammed earth construction was also used in practice in Britain, by Holland for some experimental buildings for the Duke of Bedford at Woburn; it was also used by others for villas in Winchester and other chalk districts of southern England in the middle decades of the nineteenth century[9].

In the half century before the First World War, on-site materials, including earth, were employed by a number of well-known architects. At Smeaton Manor in North Yorkshire in 1876, Philip Webb used earth extracted from the site to make the bricks for the house. Lutyens used dressed chalk at the Deanery Garden in Berkshire (1901) and even more spectacularly for Marshcourt in Wiltshire (also 1901) where it forms the main walling material. Edward Prior made rhetorical use of 'found' materials such as flint and pebbles for his 1903 house Home Place in Norfolk. In 1910 Ernest Gimson built a cottage of cob at Budleigh Salterton in Devon. This was built of sand found on site mixed with water and long straw to make walls three feet [91.4 centimetres] thick, resting on a plinth of cobble stone found in the sand[10].

These, however, were essentially one-offs. The difference with the rammed earth revival was, first, that earth construction was advocated as the official solution to the crisis of rural housing; and second, that the method promoted – *pisé de terre* – was not an indigenous regional method but an imported method initially believed to have originated in the colonies.

From the cheap cottage to the rammed earth revival

It was agreed by all involved that the revival of *pisé* in Britain was due to the efforts of one person. J St Loe Strachey was 'the revivalist of the method in England', said HO Weller in his 1922 government report[11]. According to Williams-Ellis in 1919, 'Mr Strachey himself is certainly the godfather of Pisé building as far as modern England is concerned, and his enterprise and enthusiasm are alone responsible for the present interest in the subject.'[12]

St Loe Strachey (1860–1927) was an intriguing and forceful character. The second son of Sir Edward Strachey, he was an ideologue, opinion-former, media-owner and behind-the-scenes political operator: an idiosyncratic right-winger who adopted a series of 'causes' that he promoted through, and that in turn attracted both publicity and readers to, the periodicals that he owned. Of these the most important was *The Spectator*, reportedly the most widely read political weekly of the time, which he owned and edited from 1898 to 1925; but he also owned *The County Gentleman*, a less successful title which he acquired in 1901[13]. According to Hugh Thomas, from 1898 onwards Strachey issued a stream of editorials in *The Spectator* 'providing intelligent rationalisations for conservative attitudes to the Empire.... He made a success of *The Spectator* and became quite rich in consequence'[14]. While his views were widely seen (as his daughter Amabel stated) as 'reactionary', not least because of his close association with the country landowners' lobby, his social circle was wide and heterodox, including the Webbs and Bernard Shaw[15].

Apart from the empire, the centrepiece of Strachey's political philosophy was his belief in the free market, based on the political economy of John Stuart Mill. In his autobiography of 1922 he recalled that, while a student at Balliol College Oxford in the 1880s, he had been attracted to the socialism preached by Hyndman and Henry George; but he came to the conclusion that logically, before abandoning the market system in favour of something else, first the market system had to be tried in its full unfettered form, ie 'real Free Trade'[16].

As regards the housing of the working class it was apparent to Strachey that the market was not working. In his autobiography he recalled:

> I had always been, and still am, deeply concerned in the housing question. We cannot be a really civilised nation unless we can get good houses and cheap houses for the working classes. Not being a philosopher, I had always supposed that the surest way of getting good and cheap houses was to find some improved system of construction.[17]

Three years earlier, in the introduction to Williams-Ellis' *Cottage Building in Cob, Pisé, Chalk and Clay* he made the point more directly:

> My connection with the problem of housing, and especially of rural housing, ... has been on the side of cheap material. Rightly or wrongly (I know that many great experts in building matters think quite wrongly), I have had the simplicity to believe that if you are to get cheap housing you must get it by the use of cheap material....[18]

Strachey's first venture in this regard was the Cheap Cottages Exhibition held at Letchworth Garden City in 1905. In rural areas it was the traditional responsibility of the landowner to provide housing for agricultural labourers; but by the early 1900s the cost of doing so exceeded by far what the agricultural labourers could afford to pay in rent, generally taken to be three shillings per week, which equated to a construction cost of £150[19]. The answer, Strachey believed, was to follow the precepts of Millite political economy, namely to seek a cheaper method of manufacture through technological innovation. Initially he believed that concrete was the answer and in 1904 in the pages of *The County Gentleman* he floated the idea of an exhibition of models of cheap cottages. The struggling Garden City company at Letchworth however spotted an opportunity to benefit from Strachey's publicity machine and in December 1904 suggested instead an exhibition of real cottages, which the Garden City would underwrite[20].

Strachey used his connections to establish a formidable list of supporters, headed by the Archbishop of Canterbury, and amid much publicity, the Cheap Cottages Exhibition with its 85 cottages was opened by the Duke of Devonshire in July 1905 (Fig. 69)[21]. But the winner in the £150 category, designed by Percy Houfton, was not in concrete but brick and the £150 figure was largely notional, because it excluded the cost of the site, the builder's profit and the architect's fee, as well as boundary walls, roads and sewers – and in addition the bricks for the exhibition were supplied at a special price that excluded carriage[22]. Despite its success as a publicity venture, therefore, the Cheap Cottages Exhibition failed to demonstrate either that a £150 cottage was achievable in normal circumstances or that concrete was the way to achieve it. In the subsequent cottage exhibition at Letchworth in 1907, new methods and materials were conspicuously absent and Strachey was forced to re-think his advocacy of new methods as the answer to the housing problem[23].

The problem of rural housing, however, remained and in the years leading up to the First World War it attracted increasing public attention. While willing

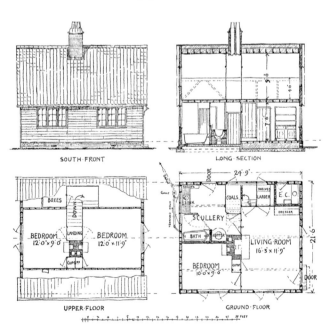

Fig. 69. FW Troup, £150 timber-framed cottage from the Cheap Cottages Exhibition at Letchworth Garden City, 1905

to subsidise the production of rural housing in Ireland under the 1906 Irish Labourers Act and giving powers to county councils in Britain to facilitate land settlement under the 1908 Small Holdings Act, with John Burns as president of the Local Government Board, the Liberal government had little to offer on rural housing[24]. This allowed the opposition Conservatives to steal the initiative, introducing a private member's bill in December 1911 that included a Treasury grant of £500,000 for rural housing[25]. By 1913 the government had responded with plans for a major rural housebuilding programme to be carried out by the Board of Agriculture, drawing on the report of a departmental committee in which Unwin, for the first time, set out his vision for state housing on a national scale [CD][26].

Strachey's contribution, characteristically, was another cheap cottage promotion. In 1913 he announced that he had built a timber-framed house for £150 on a plot at Merrow Common, close to his home at Newlands Corner in Surrey. To show off the house he organised a big opening ceremony, at which he announced a new 'challenge' to architects and other interested parties: build a cottage for £100 (later increased to 100 guineas) on land that he would supply on his estate and, if the building was still standing after a year, he would buy it[27].

Not everyone was impressed by Strachey's new campaign for cheap cottages. The rival publication *Country Life* led the attack. The 'latest thing in cheap cottages, the one put up for Mr St Loe Strachey by Mr Arnold Mitchell' for £110 was 'little better than a rabbit hutch'. To adopt this design 'as a standard ... would be to give a fatal setback to the building of adequate cottages', declared *Country Life*'s architectural editor Lawrence Weaver: beauty had to be considered as well as cost and regional traditions had to be respected[28]. In a direct rejoinder to Strachey's cheap cottage, in December 1913 *Country Life* launched a National Competition for Cottage Designs, which was to be assessed on a county-by-county basis and built by sympathetic landowners in different parts of the country, with a cost limit of £300–400 per pair, ie up to £200 per cottage. The assumption was that, as in Ireland, the portion of the rent that the rural labourer could not afford to pay would be met by a subsidy from the Treasury[29].

Like Strachey's other 'causes', the cheap cottage campaign was heavily promoted in *The Spectator*. On 22 November 1913 the magazine carried a letter from an unnamed reader in Uganda commending its efforts and drawing attention 'to a type of building, called Pisé, much used in the colonies'. The reader enclosed a cutting from a South African publication, the *Farmer's Weekly* (reprinted by Strachey), in which a certain Harold L Edwards described both the *pisé* projects he had undertaken in South Africa and the situation in New South Wales, 'where a great deal of pisé building is done'. For Strachey (who had never abandoned his interest in concrete) this new approach to building a house 'out of the stuff which is dug out of the ground' exerted an immediate appeal; and the fact that it originated (as he thought) in the colonies only added to the attraction[30]. In the introduction to the Williams-Ellis book Strachey recalled:

People who had seen and even lived in such houses wrote to *The Spectator*, and the world indeed seemed alive with Pisé de terre. I was even lent the 'Farmer's Handbook' of New South Wales [published in 1911], in which the State Government provides settlers with an elaborate description of how to build in Pisé, and how to make the necessary shuttering for doing so. It was then too that I began to hear of the seventeenth and eighteenth century buildings of Pisé in the Rhone Valley.

I had got as far as the position described above, when down swept the war upon Europe...[31]

The war however, did not mean that Strachey's interest in *pisé* came to an end. His wife Amy transformed the family home at Newlands Corner into an auxiliary hospital for troops and the resulting need for additional accommodation gave Strachey the opportunity to experiment with his new discovery. In effect his estate became a private building research station for experimenting with rammed earth. In the summer of 1915 he constructed an apple store in *pisé*, using simple shuttering he had had made on the Australian model (Fig. 70). This was followed immediately by a dining room for the patients, added on to the existing house and designed by the architect Williams-Ellis. For this, Strachey 'decided to be ambitious and experiment in '... a new form of Pisé, ie *Pisé de craie* or compressed chalk'[32]. Also in 1915, a drill hut was built for the Guildford Volunteer Training Corps using the shuttering that Strachey had constructed. He wrote to Williams-Ellis:

> Mr Swayne, an architect in the VTC, who has helped me, has made some interesting calculations. The walls, which are about 7 feet [2.1 metres] high, took a platoon, ie 52 men, 10 hours to build. The cost of 6d per hour works out, Mr Swayne tells me, at about £12.10s. He is going to make an exact calculation of what it would be in brick and corrugated iron – of course at war prices – but he thinks about £30 or £40...[33]

Fig. 70. J St L Strachey and Clough Williams-Ellis, rammed earth apple store at Newlands Corner, 1915

As we will see, exaggerated expectations of cost savings were to be characteristic of the *pisé* revival.

Some smaller structures, including a wagon house, farmyard walls (again of chalk) and a large shed, were also built in rammed earth. Characteristically, Strachey did not keep his findings to himself but used his influential connections to promote his discovery. In the 'early stages', he recalled, he was encouraged by General Sir Robert Scott-Moncrieff, who apparently issued instructions for *pisé* walling to engineer companies on the western front, based on the simple Australian shuttering design[34]. By the end of 1917 Strachey felt sufficiently confident to approach Whitehall direct. In December 1917 he submitted a proposal to the Department of Scientific and Industrial Research for them to investigate the problems arising from the use of earth and chalk for building[35].

In these wartime experiments with rammed earth, Strachey benefited from the specialist architectural input supplied by his new son-in-law, Williams-Ellis. Despite his lack of a conventional architectural training (he completed only one term at the Architectural Association before leaving to undertake his first project), Williams-Ellis built up 'quite a substantial practice, mostly concerned with country houses, large and small' before the First World War[36]. In this he was aided in part by the friendship forged with Lawrence Weaver at the time of the 1911 Gidea Park competition, which meant that he was able to share in the patronage that Weaver bestowed thanks to his position at *Country Life*[37]. In 1913 Williams-Ellis attended the opening ceremony for the £150 cottage at Merrow Common and was immediately attracted to Strachey's daughter Amabel; to ingratiate himself with the family, he entered Strachey's £100 cottage competition. His pursuit of Amabel Strachey was successful and the couple were married in July 1915, with Williams-Ellis returning from the Western Front for the wedding[38].

Rammed earth and the post-war land settlement programme

Although in the short term it brought housebuilding more or less to a halt, the main effect of the First World War was to enormously increase the political importance of housing. As well as the pledge of a general housing programme (see chapter three), the government also became increasingly committed to giving members of the armed services the direct opportunity to settle 'on the land for which they have fought'[39]. What was envisaged was a greatly expanded version of the land settlement programme instituted by the Small Holdings Act of 1908, under which between 1908 and 1914 a total of 14,389 smallholdings had been provided by county councils and county boroughs in England and Wales, including 886 with cottages[40]. A departmental committee in 1916 put forward the idea of a 'central farm' to teach settlers how to work their holdings and the Selbourne Report of 1917 proposed land settlement for ex-servicemen as part of a comprehensive policy for agriculture, including a minimum wage and minimum prices for cereals[41].

By the months following the Armistice, the land settlement programme ranked second only to the housing programme in its political importance[42]. In some ways indeed it was even more sensitive since it was targeted so precisely at active servicemen. Perhaps this was the reason that the terms of the Land Settlement (Facilities) Act of 1919 were even more generous than those of the Housing Act of the same year, with participating county councils bearing no responsibility for financial loss (unlike participating local authorities under the 1919 housing programme, who had to contribute the produce of a penny rate)[43]. Perhaps also for this reason the government decided that, while the main programme would be conducted through county and borough councils, the government would also act directly with land settlement schemes undertaken directly by the Board (later the Ministry) of Agriculture. The Board of Agriculture talked of an overall programme to settle 100,000 men within a year and in January 1919, with the troops on the Western Front increasingly restive at the slow progress of demobilisation, it issued a booklet to the troops entitled *Land Settlement in the Mother Country*. This summarised the scheme and stated that 'any man who desires to obtain, after demobilisation, a Small Holding of not more than 50 acres [20.2 hectares] in England or Wales should fill in the form printed in the middle of the booklet'[44].

The person appointed to take charge of this politically charged programme was Lawrence Weaver. In 1916 Weaver had left *Country Life* to join the reserves and was then transferred to the Food Production Department – one of the Whitehall success stories of the war – where he proved an administrative 'star', becoming Controller of Supplies and being rewarded in 1918 with the CBE[45]. In December 1918 Weaver was appointed commercial secretary of the Board of Agriculture at a salary of £2000, higher than that of the permanent secretary and the same as that of the president of the Board[46]. As such he was responsible for implementing the land settlement programme and what was effectively the national rural housing programme embedded within it.

As regards the kind of houses to be provided on the new smallholdings, under Weaver the Board was 'in complete sympathy with the new attitude towards housing matters in this country, which was manifested immediately after the Armistice'– in other words, the recommendations for a substantial improvement in housing standards made by the Tudor Walters Report [CD][47]. Reversing the policy followed at its war-time development at Patrickton in Yorkshire, with Weaver in post the Board declared that it wanted the houses built for smallholders to be of exemplary character. All were to have a parlour in addition to a living room and scullery and (despite ridicule from some quarters) all were to have a bathroom[48]. But where the Tudor Walters Report and, following it, the Ministry of Health looked to 'standardisation and simplification', for the smallholdings programme the Board of Agriculture placed its faith in 'using, as far as possible, local materials and traditional methods of construction'[49]. Both the belief in 'good' rather than 'cheap' cottages and the idea of promoting local traditions in design and construction were consistent with Weaver's pre-war position at *Country Life*.

Prominent among the methods of construction promoted by the Board of Agriculture was rammed earth. Weaver's department was 'inclined to plume itself on its early appreciation of the potentialities of pisé'[50]. Following Weaver's appointment, it was decided that the first post-war development – at Amesbury in Wiltshire – should act as a flagship for the programme, not just providing housing in 'a rural district on the lines of the Report of the Committee presided over by Sir Tudor Walters' but also undertaking 'experiments in the use of local and special materials and methods of construction, at a time when the cost of accepted methods and materials was extremely high'[51]. The plan for the Amesbury settlement was to erect a number of cottages in a variety of raw earth methods and compare them, not just against each other, but also against cottages built at the site in concrete (of various sorts) and timber, as well as conventional brick construction.

The Department of Scientific and Industrial Research (DSIR) was also invited to take part in the Amesbury experiment. On his appointment as commercial secretary in December 1918, Weaver proposed to the DSIR that it too should erect some experimental cottages at Amesbury, either with local materials using methods which had fallen into disuse, or by new methods[52]. The DSIR was headed by another prominent member of the country landowners' lobby, Lord Curzon, who at this stage was blocking the request for a Building Research Board to be established to conduct the research needed for the housing programme (see chapter ten). In contrast, Weaver's proposal for research into rural methods at Amesbury was accepted immediately[53]. At the instigation of the Board of Agriculture, the pioneer of reinforced concrete Alban Scott – the source for much of the information being collated by Williams-Ellis for his book – was appointed as consulting architect for the DSIR scheme while WR Jaggard – best known as co-author with FE Drury of *Architectural Building Construction* (1916–) – was appointed as architect in charge[54].

If county councils were to build houses of the kind approved by the Board, information and model plans would be needed. A circular letter issued by the Board on 18 December 1918 urged councils to 'proceed at once' with land settlement schemes for ex-servicemen and informed them that the Board would provide them with 'all possible assistance in regard to the design and plan [*sic*] of suitable cottages and buildings for Small Holdings'[55]. This meant putting in place an administrative structure with a regional tier of district commissioners (similar to the housing commissioners appointed under the housing programme) and a team of architects in Whitehall who could produce model designs suitable for different conditions and requirements[56].

First of the superintending architects appointed at Whitehall was Williams-Ellis. In his 1933 memoir of Lawrence Weaver, Williams-Ellis gave a characteristically colourful (if not wholly accurate) account of how in late 1918, with the Armistice approaching, he got Weaver to request his early release so that he could join the Board of Agriculture. The result was that 'within a fortnight of the Armistice I was actually back in London in the guise of a "Pivotal Man" … urgently needed by the Ministry of Agriculture for the furtherance of its small-holdings and land-settlement schemes.' He continued:

> It was over the department concerned with such matters that Lawrence now reigned … and there, very soon, were congregated a little band of ex-soldier architects, all old friends of his, all devoted to him personally and now filled with his own enthusiasms for a new and better physical England, and all, if one of them may say so, well chosen for the work in hand.
>
> Anyway, there we were, Maxwell Ayrton, Oswald Milne, Hugh Maule, and the rest, back at our drawing-boards and all turning out plans of jolly little houses and farm-buildings for small-holders appropriate to this and that acreage and type of family and to this or that part of the British Isles. Jointly, and always under Lawrence's steadying and realistic leadership, we quickly produced an imposing corpus of work – plans, details and specifications – which we condensed into a Government publication that became, as it were, a trade catalogue of the wares and ideas that we had to offer[57].

Williams-Ellis joined the Board as a superintending architect in January 1919 (two months rather than two weeks after the Armistice), followed in March by HPG Maule and OM Ayrton and, as assistant architects, FWJ Hart, T Tyrwhitt and HPR Aitchison[58]. The 'trade catalogue'– the *Manual for the Guidance of County Councils and their Architects in the Equipment of Small Holdings: Part I: The Planning and Construction of Cottages* – was issued in May.

Williams-Ellis' period in Weaver's department was short; he says 'three months' but it was more like six (which was still not long given that he had used it to get early release) and he left in the summer of 1919. By then he had virtually completed his compendium on earth construction, *Cottage Building in Cob, Pisé, Chalk and Clay: A Renaissance*, which was published by *Country Life* in the autumn with a substantial introduction by St Loe Strachey. Freely acknowledging his debt to those whose work he exploited (and often quoted at length) including Strachey for *pisé*, a Mr Fulford of Devon for cob and above all Alban Scott for the 'mass of laboriously collected and carefully arranged information' that 'made this book possible at all',[59] Williams-Ellis advanced the case for earth construction with the fervour of an apostle. Given the severe shortages of labour, materials and transport, he wrote:

> So far as rural housing is concerned, the solution must be sought through the use of natural materials already existing on the site…. It is not so much a question as to whether a Cob or Pisé house is preferable to one of brick or stone or concrete … but as to whether you will boldly revert to these old and well-tried methods of building, or … build nothing at all.[60]

Fig. 71. Clough Williams-Ellis, single-storey pisé cottage at Newlands Corner, 1919

Before proceeding to build any rammed earth cottages at Amesbury, the Board of Agriculture decided that a prototype should be constructed. This was achieved in collaboration with Strachey, who in the summer of 1919 erected at Newlands Corner a single-storey three-bedroom parlour cottage in *pisé* (Fig. 71). The shuttering was designed by Williams-Ellis and constructed by the Board, the plan was from the Board's *Manual* (Type A) and supervision was provided by Williams-Ellis on behalf of the Board (Fig. 72). At the end of August the building was inspected by a team from the Board, including Ayrton and Tyrwhitt, who in a report dated 2 September enthused that, 'The results of the experiment have been entirely satisfactory.' The entire cottage, excluding foundation and base, had taken only 400 man-hours to erect, equivalent to a cost of £20 for the walls – and this at a time when the average tender price under the housing programme was £740[61].

Following completion of the Newlands cottage, the shuttering was sent to Amesbury, so that tests could be carried out with the Amesbury soil before starting construction of what was described by the Ministry of Agriculture (incorrectly, as we now know) as 'the first two-storied [sic] pisé dwelling erected in England' (Figs 73 and 74)[62]. Work on the 2500 acre [1011.7 hectares] site (the purchase of which was finally agreed in April 1919)[63] started in May 1919, with construction of the office (for the Board's staff), hostel (for the building workers) and road (Fig. 75). Responsibility at Whitehall for construction rested with superintending architect Thomas Tyrwhitt[64]. By November 1920, 25 out of the Ministry's 27 houses were complete and the other two, plus the DSIR's five, were finishing[65]. Of the 32 houses, six were *pisé* (four in *pisé de terre*, one in rammed chalk and one in rammed chalk-cement); two were cob, the argument being that, although by no means indigenous to Amesbury, this method should be tested as well; three were timber (two timber frame and two re-used army hostels); four were concrete of various sorts, including monolithic and concrete block; and the remainder were brick[66].

144

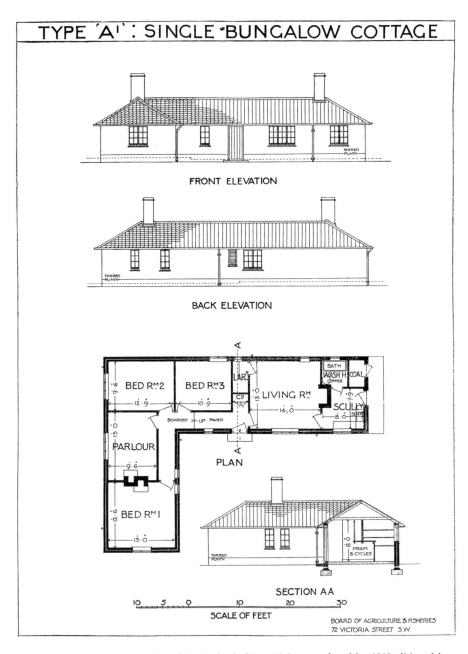

Fig. 72. Clough Williams-Ellis, plan of the Newlands Corner pisé cottage, from May 1919 edition of the Board of Agriculture's Manual

Fig. 73. The first pisé cottage at Amesbury built by the Board of Agriculture, 1919–20

Fig. 74. Plinth detail of the first pisé cottage at Amesbury

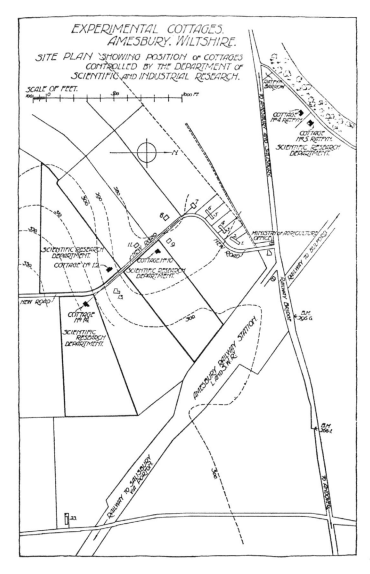

Fig. 75. Part-plan of the Amesbury settlement showing location of the five DSIR cottages

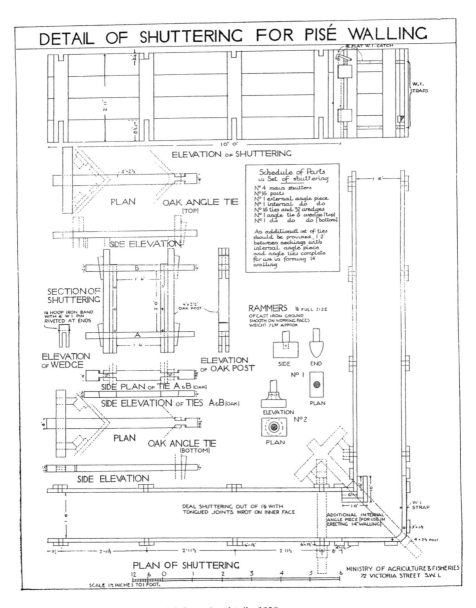

Fig. 76. Ministry of Agriculture, pisé shuttering details, 1920

By June 1920 the Ministry felt that enough had been learned from the Amesbury experiment to publicise the results. The site was opened one day per week for visits by interested parties[67]. In September an interim report was issued, both in the Ministry's journal and as part of a new edition of the *Manual*. This declared that although not final or complete, 'the data already obtained are sufficiently definite for pisé construction to be embarked upon with satisfactory results'[68]. One caveat, learned from experience of the first cottage, which

Fig 77. The pisé pair at
Amesbury, 1920 (left),
street front, and Fig. 78
(below), plans, section
and elevations

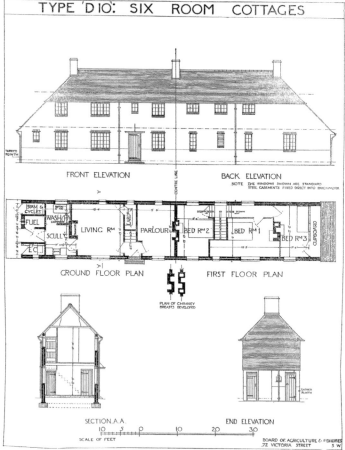

TYPE 'D 10: SIX ROOM COTTAGES

FRONT ELEVATION

BACK ELEVATION

NOTE THE WINDOWS SHOWN ARE STANDARD STEEL CASEMENTS FIXED DIRECT INTO BRICKWORK

GROUND FLOOR PLAN

FIRST FLOOR PLAN

PLAN OF CHIMNEY BREASTS DEVELOPED

SECTION.A.A.

END ELEVATION

SCALE OF FEET

BOARD OF AGRICULTURE & FISHERIES
72 VICTORIA STREET   S W

started in the autumn of 1919, was that construction of *pisé* walls during winter
should be avoided; if the earth became wet, ramming could not be carried out
satisfactorily, with consequent waste of time and money. The best shuttering
to use was the simple wooden form designed by the Ministry (Fig. 76), rather

than Williams-Ellis' earlier, more complicated design. But, said the Ministry, if properly conducted, rammed earth offered substantial cost savings. The realised cost for the *pisé* pair, with the usual 18 inch [45.7 centimetres] walls to the ground floor and 14 inch [35.5 centimetres] to the first floor, was said to be 15 shillings per yard super as against 25 shillings for 11 inch (27.9 centimetre) cavity brick walls – a saving of 40% (Figs 77 and 78)[69].

Lessons from the Amesbury experiment

The advocacy of 'Building in Pisé de Terre' in the *Manual* of September 1920 can be seen as the highpoint of the *pisé* revival. Little more than six months later, in April 1921, Weaver gave a lecture at the RIBA entitled 'Building for Land Settlement: A Survey of the Ministry's Work', in which advocacy of *pisé* was conspicuous by its absence. With admirable *sang froid*, Weaver simply observed that, 'In building operations it has been found that brick has held its own, though most exhaustive experiments have been made with cob, pisé and concrete.'[70]

What lay behind this change of view? In part it was the general change in the economics of building that took place in the winter of 1920–21. With the sudden collapse of the post-war boom, traditional materials and labour again became available and building costs started to fall from the 'monopoly' levels they had reached in 1919 and 1920. In other words, the crisis that the *pisé* revival had been designed to overcome no longer existed. There was also the reversal in 1920–21 in the attitude of the government towards the social programmes it had instituted in the aftermath of the Armistice, as what had seemed at the time prudent measures required to honour pledges to the returning 'heroes' took on the appearance of reckless extravagance (see chapter three).

But more immediately there were the cost conclusions from the Amesbury experiment. Notwithstanding the statements made in the Ministry's interim report, the earth methods spectacularly failed to deliver the cost savings which the *pisé* revivalists (from Strachey, to Williams-Ellis, to the Ministry itself) had claimed. With its primitive technology, the economic viability of rammed earth depended heavily on the plentiful supply of cheap labour. In post-war Amesbury this was simply unavailable; building workers had to be brought nine miles [14.4 kilometres] from Salisbury and accommodated on site, 'thus adding very considerably to the cost of the works', as Jaggard noted[71].

A memorandum dated 30 September 1920 set out the stark facts. Even setting aside the first cob and *pisé* cottages, which cost £1495 and £1304 respectively, the *pisé* cottage then finishing was coming in at £883, as against almost the same amount (£889) for a brick pair. Even the *pisé* pair, which in the end was by far the most economical of the *pisé* structures, came in at £1459, more than 60% more than the brick pair[72].

A subsequent analysis of expenditure on all 32 cottages at Amesbury underlined the point. Taking the materials by type (*pisé*, cob, concrete, timber-framed and brick), the cost hierarchy was almost exactly the inverse of what the *pisé* revivalists at the Ministry had claimed. Concrete came out the cheapest, with an average cost for the four cottages of £1284. Next came timber framing (both new build and converted huts), at £1395. For the 16 traditional brick cottages the average figure was £1532. Next came *pisé*, at an average of £1885 for the six cottages, a figure that was topped only by cob, an average of £2281 for the two cottages[73]. However great its enthusiasm for earth materials, the Ministry had little choice but to accept that its experiments had shown that *pisé* was not economically viable for rural housing.

The DSIR echoed the conclusion. Its annual report for 1920–21 (dated August 1921) noted the completion of the Amesbury scheme during the year and commented:

> The only general conclusion it is safe to draw from the experiment confirms that of the past year's experience over the kingdom, that substantial, economical progress is to be sought neither in blind reversion to ancient practice nor in the hasty adoption of revolutionary methods, but by steady scientific development of the normal.[74]

Technically, the most positive outcome of the Amesbury work was seen as the success of the 'chalk pisé' method developed by the DSIR's consulting architect Alban Scott, which combined chalk and Portland cement in the ratio of 20:1 (Figs 79 and 80)[75]. In a report on Amesbury published in the Ministry's *Journal* in September 1920, Williams-Ellis noted that for cob 'the cost was discouraging, but the chalk and cement method is distinctly promising'[76]. The Building Research Board (BRB) which, when finally established in 1920, took over responsibility for the DSIR cottages at Amesbury, took the same view, its director of research HO Weller telling the board in November 1920 that 'the outstanding justification for the expenditure, so far, was the chalk cement walling'[77]. He took the same view in his (less than enthusiastic) report on *Building in Cob and Pisé de Terre*, published by the BRB in 1922, suggesting that the future of rammed earth, such as it was, lay in combining 'pisé de terre with cement concrete.... to the benefit of both materials'[78].

The story with the land settlement programme overall was rather more positive. Unlike the 'homes fit for heroes' programme, the land settlement programme survived the Treasury-led cutbacks of 1920–21. In the summer of 1920 the cabinet imposed limits to both the capital cost and annual loss per smallholding, leading to a reduction in space standards, as set out in the September 1920 *Manual*[79]; but the programme itself survived. The architectural work at Whitehall, however, lost its allure and became largely routine. There were no new editions of the *Manual* after September 1920; the 'great majority' of the post-war holdings were occupied by 1921[80] and after May 1921 the design of houses for smallholdings no longer figured in the pages of the Ministry's *Journal*. By that year, of the original six senior architects at Whitehall – Williams-Ellis, Maule, Ayrton, Hart, Tyrrwhitt

Fig. 79. Ramming of chalk-cement walls at Amesbury

Fig. 80. DSIR cottage in rammed chalk-cement at Amesbury, 1920

and Aitchison – only Maule remained and at the end of 1922 Weaver himself resigned, to head up the British Empire Exhibition at Wembley[81].

The land settlement programme was effectively completed by 1926. Figures given by the Ministry in October 1925 showed that 16,461 ex-servicemen and 2221 civilians were occupying post-war smallholdings, a total of 18,682. Up to the end of 1924 county and borough councils had built 2749 houses and it was estimated that the total cost of the building programme would be some £5.5 million. It was also estimated that the total capital expenditure on the programme in England and Wales would be £16 million, of which half would be written off by the Treasury[82]. While this was a significant achievement, critics might observe that it hardly eclipsed the 14,400 smallholdings and 866 cottages achieved between 1908 and 1914 without either a Treasury grant or the elaborate administrative structure established by the Ministry.

Arguments for rammed earth, past and present

In the arguments advanced for rammed earth in the period 1905–1925, a number of distinct strands can be identified. First and most crudely, there was the search for a cheap material that would deliver what otherwise appeared unattainable – a cottage that the rural labourer could afford. This was identified most clearly with Strachey, whose search for a cheap cottage at Letchworth and Merrow before the war led directly into the war-time experiments with *pisé*. In Strachey's case this derived from a right-wing political philosophy that saw the market as holding the solution to social problems and which celebrated what were seen to be the colonial origins of rammed earth. Yet in essence, the search for a technological route to cost reductions was one that would arise almost whenever social democratic governments sought to undertake construction programmes for social ends; in the case of Britain this extends from the well-known experiments with steel and concrete in the 1920s (see chapter 11) and the 1940s to the endorsement of 'modular', ie prefabricated, construction in the New Labour government of Tony Blair[83]. So far however (notwithstanding the Bath/DTI study) it has not been adduced in the arguments for the present-day revival of rammed earth, for the simple reason that in advanced capitalist economies – where technology is readily available and labour is expensive – rammed earth is not particularly cheap.

Second, in the 1905–1925 revival there was the belief in local traditions in architecture: the idea that architecture should work with the materials available to hand in the locality, with the use of earth materials found on site being seen as the ultimate in this approach. The 1914 *Country Life* competition for cottage designs devised by Lawrence Weaver was the clearest expression of this commitment to local materials and methods, which from December 1918 became the official policy of the Board of Agriculture. There were two aspects to this position: what one might term a rational side, which accepts that, other things being equal, it 'makes sense' to use materials already available in the locality; and a romantic side, which sees the use of local materials as a rhetorical device and particularly as a protest against the universalising tendencies of modernity. Both of these aspects were present in the arts and crafts commitment to local materials and as such ran through into the modern architecture of the twentieth century, from Ernst May's explorations of a modern vernacular in 1920s Silesia to Kenneth Frampton's call for 'critical regionalism' in the 1980s[84]. For obvious reasons, the cogency of the 'rational' aspect was greatly increased if one was building in remote areas or at times when transport and conventional materials were unavailable; this the *pisé* revivalists imagined would be the case in post-1918 Britain, as did the US Department of Agriculture in rural America in the Great Depression of the 1930s[85]. But in relation to the rammed earth revival of 1905–1925, there was always a problem in arguing for rammed earth in regionalist terms. While the material, as at Amesbury, was indisputably local, the method was not.

The third strand was the belief in rammed earth as a modern material. Indicated by the title of Williams-Ellis' 1920 article, 'The Modern Cottage: Experiments in Pisé at Amesbury'[86], this was the position of Alban Scott and the DSIR, epitomised by Scott's development of rammed chalk-cement. By the application of science, it was believed, rammed earth might become a modern material to match or exceed any other in terms of performance, economy and comfort. This was perhaps the most radical vision, for it saw rammed earth not as a material of the past but of the future and, instead of valuing its regional character, embraced its universality. The DSIR saw the combination of chalk and cement as the future of earth materials and in a sense they were right, at least as far as the next 50 years was concerned, for in the colonial context of the 1940s some of the most fruitful applications of earth materials involved mixing them with Portland cement to produce 'stabilised earth'. Thus when Williams-Ellis' book was revised and reissued after the Second World War, the title was changed to *Building in cob, pisé and stabilized earth*, a recognition of the role that cement-enriched mixtures now played in rammed earth[87]. This conception of rammed earth, we might note, is entirely alien to present-day revivalists such as Rowland Keable, for whom a key attraction of rammed earth is precisely that it offers an eco-friendly alternative to concrete. Indeed, one of the lessons from the events of 1905–1925 is that far from representing opposed positions, as they often appear to do in contemporary thought, concrete and rammed earth were seen as having much in common, being simply different techniques for turning earth into a usable and useful constructional material.

So of these three strands in the argument for the revival of rammed earth in 1905–1925, which if any do we find in the European rammed earth re-revival of today? Rammed earth is promoted today not because it is cheap, nor because conventional materials are not available, nor because it can be combined with cement. Effectively, the only argument from the early twentieth-century revival that we find prominent today is the 'romantic' element of Weaver's regionalist position: the idea that, against the universalising tendency of modernity – in the nineteenth century, it was the national market for building materials; today, it is 'globalisation' – there is a moral obligation to stand up for what is specific to a place and a region. The main argument adduced for the rammed earth today by proponents such as Rowland Keable is primarily ecological, stemming from the transformed outlook on the planet and its resources brought about by the environmental movement: rammed earth uses less energy to produce and transport than almost any other material. In some cases (notably Martin Rauch) this ecological argument is overlaid with an aesthetic appreciation of the special visual and sensory qualities that can be offered by high-quality rammed earth. Neither of these arguments, the ecological nor the sensory, was a factor in the rammed earth revival of 1905–1925.

Chapter ten

Breeze blocks and Bolshevism

The Building Research Station (BRS) is widely recognised as the first government organisation in the world dedicated to research in construction. First established in 1921 at Acton in west London (Fig. 81), the BRS created a model that was subsequently adopted across the globe. As the president of the Conseil International du Bâtiment (CIB), G Blachère, put it on the occasion of the 50th anniversary of the BRS in 1971, the BRS provided a model for state-funded building research that was followed 'in the Commonwealth, in Europe and now all over the world'[1].

In outline, the story of how the BRS – later known as the Building Research Establishment (BRE) – came into being is well known. As the government became increasingly committed to a major post-war housebuilding programme at a time when both materials and labour were in short supply, it became evident that non-traditional methods and materials would be needed and therefore that research was required to establish which were suitable. But whereas the process by which the state became involved in a programme to build 500,000 houses has been extensively investigated by historians using unpublished government

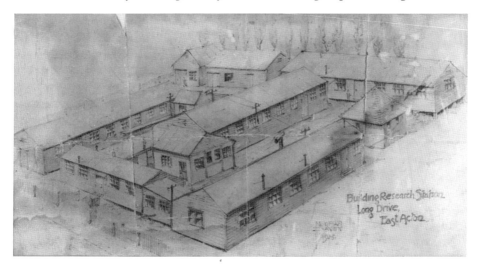

Fig. 81. The Building Research Station at Acton, drawing by Norman Davey made at the time of the move to Garston, 1925

papers (Johnson 1968, Wilding 1970, Gilbert 1970, Swenarton 1981, Fraser 1996)[2], the process by which it became involved in building research has not been subjected to the same kind of historical scrutiny. Instead, the existing accounts of the origins of the BRS (many of them devised to mark the 50th anniversary of the BRS in 1971) have either been produced by the BRS itself (White 1965, White 1966, Lea 1971, Building Research Station 1972) or by those personally involved in its formation or activities (Heath and Hetherington 1946, Atkinson 1971) – and the best of these, the 1966 history by RB White, has remained unpublished hitherto [CD][3].

Perhaps inevitably, retrospective accounts of this sort tend to smooth over the conflicts and controversies involved in major historical developments of this kind and suggest a kind of inevitability to the outcome, as though the introduction of a government organisation for building research after the First World War was a foregone conclusion. But if we go behind the scenes and look at unpublished government papers, we see that on the contrary the proposal for government-funded building research was from the start a highly controversial innovation, the adoption and implementation of which was by no means assured. Championed by those who believed that the war had permanently changed the relations between social classes and that a new 'social contract' was required from the government, building research was equally vehemently opposed by those who believed that after the war, things should go back to 'normal' (ie as they had been before the war) or something close to it – and therefore that there was no need for the government to undertake building research.

The Building Materials Research Committee 1917–20

As shown in chapter three, in the summer of 1917 the cabinet faced what appeared to be a major crisis in the prosecution of the war, with engineering workers on strike and widespread industrial unrest. A commission of enquiry was set up in June 1917 which reported that a significant cause of industrial unrest was the housing shortage and recommended that, if nothing else, 'announcements should be made of policy as regards housing'[4]. Accordingly on 24 July 1917 the cabinet authorised an announcement of the government's intentions for post-war housing and two days later the president of the Local Government Board (LGB) appointed a committee to investigate the questions of building construction that would arise[5].

Following its formation in July 1917, it was immediately evident to Unwin and others and the Tudor Walters committee that the resources of traditional building would not be sufficient for a major post-war housebuilding programme; and therefore, if untried alternative methods were to be used as a supplement, research and experiment were needed to establish which ones were safe and satisfactory in performance[6]. During September 1917 the materials sub-committee, which was chaired by Unwin, discussed the problem and reached the conclusion that for the post-war housing programme:

Timber will not be available except to a very limited extent. New methods and materials of construction must, therefore, be sought, and these can only be adopted with advantage after careful test and experiment.[7]

Tests were needed, they said, on the use of concrete in place of timber for floors and roofs of cottages, and on the penetration of air and moisture through walls of concrete compared with those of brick. In addition, they wanted investigation into the possibilities of increasing the output of timber and brick by such means as new processes in brick-making and the artificial seasoning and sterilising of timber[8].

Already, on 30 August 1917, Unwin had contacted the Department of Scientific and Industrial Research (DSIR), the department established the previous year to direct government research, and asked:

What procedure his committee should adopt in order to instigate some research into building materials generally, but with special reference to the use of concrete as a substitute for brick or wood in the construction of cottages.[9]

At this stage, the request for building research had the support of the LGB and accordingly on 3 October the advisory council of the DSIR agreed to the formation of a Building Materials Research Committee (BMRC) to undertake the research required by the Tudor Walters committee[10]. To ensure close liaison between the two committees, it was decided that Unwin would be chairman of the BMRC and E Leonard, the secretary to the Tudor Walters committee, would be its secretary[11]. Tudor Walters was told by the secretary of the DSIR, Sir Frank Heath, that as chairman Unwin would direct the researches and that the 'Committee would only meet when the chairman thought it necessary to consult his colleagues'[12]. The other members of the BMRC were the chief engineer at the London County Council (LCC), GW Humphreys; the architect ES Prior, as representative of the Royal Institute of British Architects; and Seebohm Rowntree, who was already involved with research on floor coverings at his York factory and was, as Beatrice Webb noted, 'eager to spend his time and money' on housing questions[13].

The recommendation for the establishment of the BMRC had been achieved with comparative ease. Thereafter, however, the progress of research into new building methods was rather less smooth than this first step suggested, and was a good deal more problematic than the published accounts suggest[14]. The advisory council on 3 October 1917 had recommended only the formation of the BMRC; its scope and, most importantly, its financial allocation had still to be settled. Here a major difference of view soon emerged, based on the two very different conceptions of government policy, and specifically post-war housing, then current in Whitehall, as represented primarily by the Ministry of Reconstruction and the LGB[15].

The Tudor Walters committee, like the Ministry of Reconstruction, believed that central government should take responsibility for the post-war housing programme and should ensure, by whatever means might be necessary, the building of some 300,000 houses within two years after the war. In contrast, the LGB preferred to see post-war housing policy as a development of, rather than a departure from, pre-war traditions; housing would remain the responsibility of local authorities and the role of central government would be merely to offer financial assistance. The view taken by the Tudor Walters committee implied that resources should be made to match the requirements of the housing programme, whereas the view of the LGB suggested the opposite. In the first case, new methods of building would play a major role in the housing programme and building research would be essential to decide which methods should be used. In the latter case, new methods would be used only where and if they reduced the cost of building, and research would be confined to those methods that could be expected to reduce costs.

These opposing views reflected a wider division within the government. In the months following the Armistice Lloyd George finally persuaded the cabinet to adopt the housing policy associated with the Ministry of Reconstruction; as the government spokesman told MPs, 'the money we are going to spend on housing is an insurance against Bolshevism and Revolution'[16]. But, even so, many influential interests, both outside and within government, remained opposed to the new programme, which was regarded by the City and the Treasury as a costly extravagance antithetical to the holy grail of 'sound finance' (see chapter three).

In relation to building research, these differences of opinion first emerged in January 1918. In response to requests received from the Tudor Walters committee, on 3 January 1918 the BMRC submitted to the advisory council of the DSIR a report in which it proposed a wide-ranging programme of research into building materials. The proposal covered three main areas: experiments in constructional work, mainly involving substitutes for timber in walls, floors and roofs; tests on timber, including artificial seasoning and tests on specimens from alternative sources of supply; and research into the 'effect on comfort and health' of certain new methods, primarily tests on the water resistance and thermal conductivity of various forms of concrete walls. In all, the report proposed 12 different researches and its estimate of the cost of the programme (£5740) included a liberal allowance of £1200 for researches still to be specified. Although the total might seem a lot, the BMRC wished 'to point out that if, as a result of the suggested researches, a saving of only £1 can be effected in the cost of construction of each of the cottages contemplated under the housing scheme, the saving to the nation would be £300,000'[17].

The response of the LGB came in a letter to the secretary of the DSIR dated 9 January. The LGB raised two questions in regard to the BMRC's report: whether previous experience or experiments had not already answered the questions proposed for research; and whether 'the new methods and forms of construction will in fact spell economy in building'[18]. In view of this comment from the sponsoring department, the advisory council of the DSIR could scarcely do other than refer back the BMRC report. The programme proposed, the DSIR noted, 'will extend over a longer period and involve a larger expenditure than was proposed by the [Advisory] Council in the first instance'[19]. Ostensibly to strengthen the committee, but in reality to put a check on what the Office of Works called its 'unnecessary and extravagant' demands[20], representatives of the Office of Works and the LGB were added to its membership. These were RJ Allison, the principal architect at the Office of Works, who had been responsible for the 1917 housing scheme for the Royal Aircraft Factory at Farnborough; and PM Crosthwaite, an engineering inspector at the LGB[21]. Instead of approving the research programme as a whole, the DSIR decided to appoint two technical officers, EH Tabor (an engineer with the London County Council) and Hugh Davies (an inspector with the Board of Education), to review and report on each item in the programme, reserving to itself the right to approve or reject each item on its merits. The function of this was revealed by DSIR chief Frank Heath: 'As each section of work is reviewed by these two officers, it comes up to the advisory council for approval.... By proceeding in this manner, the Department will only be committed to particular investigations'[22].

The result was that the researches undertaken by the BMRC were considerably fewer in number, and took considerably longer to complete, than had originally been envisaged by the committee. Approval was eventually obtained for expenditure on six main investigations. Researches into timberless floors (reinforced concrete, hollow brick etc) and the stability of thin walls were approved by the advisory council in March 1918. These were followed in June by investigations into the transmission of heat and gases through, and the condensation of moisture on, walls of concrete and brick and into new kinds of cooking ranges[23]. In addition, later in 1918, approval was secured for two other projects related to the housing programme: the use of slag and breeze as aggregates for concrete and the properties of lime mortar as an alternative to Portland cement[24].

These investigations eventually produced useful results, published as *Special Reports* by the Building Research Board [CD]. The tests on floors, for instance, showed that composite floors of brick, tile or concrete reinforced with steel bars performed satisfactorily under the loads found on the upper floors of small houses and could be regarded as a suitable alternative to floors of timber. The test on walls showed that walls of concrete blocks had a greater resistance to crushing than those of brick but that their thermal performance was inadequate unless a cavity was provided and one of the skins was of coke breeze[25].

These were results of obvious relevance to the housing programme launched by the Addison Act of 1919. But, due in large part to the delays arising from the procedure adopted for the BMRC, they were not made available in time to be taken into account in selecting new methods for use in the housing programme. In May 1920 the Ministry of Health (the successor to the LGB, set up by Lloyd George in 1919 to deliver the 'homes fit for heroes' programme) complained that no report of the findings of the BMRC had been made available[26]; and it was not until July 1920 that, at the instigation of the Ministry, the first results of the BMRC's investigations (on walls) were published, in the Ministry's journal *Housing*[27]. By this date, under pressure from the Ministry, most of the large municipalities had already entered into contracts with various firms for houses built using new methods and/or materials[28].

The Building Research Board 1919–21

At the beginning of 1919, with the adoption by the government of the housing campaign, the advisory council endorsed a proposal from the BMRC [CD] calling for a permanent body to be established to conduct building research[29]. The recommendation, however, was rejected by Lord Curzon who, as Lord President of the Privy Council, was the minister with responsibility for the DSIR[30]. Instead, Curzon gave his approval to a rival proposal much closer to his heart – the revival of traditional techniques of earth construction proposed by the Board of Agriculture for the small holdings programme (see chapter nine)[31].

Accordingly, it was only in the autumn of 1919, when the cabinet realised that the shortage of bricks and bricklayers was threatening the entire housing programme, that renewed action was taken to secure research into new methods and materials. On 27 October 1919 the minister of health, Christopher Addison, submitted a cabinet memorandum in which the use of new methods ranked high among the proposals for 'drastic action' to address the problems faced by the housing programme[32]. While rejecting Addison's call for an effective system of building controls, which was vehemently opposed by private enterprise and the City as well as the Treasury, the cabinet was happy to endorse the wider use of new methods of construction. Research was therefore needed to establish which methods were satisfactory and at the beginning of November 1919 Addison asked whether the DSIR was prepared to undertake this[33]. At this date, with the BMRC concluding its programme of research, there was no organisation at the DSIR capable of conducting the work requested by the Ministry of Health. Accordingly on 17 December, the advisory council renewed its recommendation for the establishment of a Building Research Board[34]. By this date Curzon had been succeeded as Lord President by AJ Balfour, one of the leading advocates of industrial research, and this time the recommendation was accepted.

Although the DSIR had agreed to the establishment of a Building Research Board (BRB), progress in its formation was, in view of the urgency of the request from the Ministry of Health, dilatory in the extreme. In February 1920 the Department decided to 'go slowly with the establishment of the Board and to build it up as suggestions for really good men were secured'[35]. As a result the BRB was not constituted until June 1920. In the absence of a single person combining technical qualifications with a 'general reputation', it was decided to separate the functions of chairman of the Board and director of research. Lord Salisbury, an acknowledged authority on housing and a person of 'wide outlook' and national standing, agreed to become chairman and HO Weller, an engineer seconded from the Indian Civil Service, was appointed as director on a temporary basis[36]. On its formation the BRB took over the DSIR's existing activities in building research, including the experimental earth cottages at Amesbury and also the BMRC, which was finally wound up in December 1920[37].

The selection of the members of the Board proved tortuous. The DSIR proposed Sir Aston Webb (representing the architectural profession), Major-General Heath (formerly engineer-in-chief to the British armies in France), and GW Humphreys (the LCC chief engineer and a member of the BMRC), all of whom agreed to serve, plus representatives of the Ministry of Health and Office of Works. As its representative the Ministry of Health nominated SB Russell, Unwin's former assistant at Gretna and now chief architect, alongside Unwin, at the Ministry. As regards the Office of Works representative, the strong preference at the DSIR was for RJ Allison (chief architect under Sir Frank Baines at the Office of Works and a member of the BMRC) rather than Baines, who was regarded as something of a contrarian; as one senior official put it, 'I trust there may be no doubt about this nomination, as if Sir Frank Baines were nominated I think the whole business might just as well stop at the present moment'[38]. But when the Office of Works was approached by the DSIR about a nomination, the response was frosty. Under the mercurial entrepreneur Sir Alfred Mond, the Office of Works had pushed repeatedly but unsuccessfully to be given the lead role in Whitehall for the housing programme[39]; and the department was by no means pleased at this latest incursion into what it regarded as its rightful territory. The DSIR was informed in June 1920 that the Office of Works 'did not think much' of the request to nominate a representative on the board; 'their own Department had conducted a great deal of research work on the subjects concerned during the war … they had got many valuable results and they were still carrying on such work'; and furthermore the appointment of a representative to the BRB would take up the time of an already busy officer[40]. In the end Mond relented, albeit 'only with some hesitation and reserve'; he was 'somewhat surprised that this Department, as the Government Building Department, should not have been consulted either as to the desirability of establishing the Board or to its membership'[41] and, in a deliberate snub, nominated neither Baines nor Allison but a staff architect, AR Myers. Only when the status of the departmental representatives was elevated from associate members to full members did the Office of Works agree to Allison replacing Myers as its representative on the board[42].

Until the latter part of 1921 the BRB concentrated almost entirely on questions relating to cottage construction for the housing programme. The preservation of stone, largely in connection with Lethaby's work on Westminster Abbey, was almost the only subject not related to housing in which the board took an interest at this stage[43]. But even on questions connected with the housing programme, work on research was considerably delayed. Rather than depending on existing establishments such as the LCC School of Building at Brixton or the National Physical Laboratory, as the BMRC had done, it was decided that the BRB should have its own experimental station (although, as Salisbury told the Board, to meet Treasury demands for economy it was decreed that 'no more accommodation should be provided in the first place than was absolutely necessary')[44]. In the autumn of 1920 the Board took over from the Ministry of Health a site at Acton in west London for use as a Building Research Station and early in October 1920 the Board notified the Office of Works of its requirements in terms of accommodation. But by February 1921 nothing had happened and so it was decided that a contractor should be engaged instead; Sir James Carmichael, the former director general of housing at the Ministry of Health, offered the services of his firm, working on a prime costs basis[45]. As a result it was not until July 1921, just as the Addison Act housing programme was being axed by the cabinet[46] that the BRS's modest suite of buildings at Acton – timber huts containing small

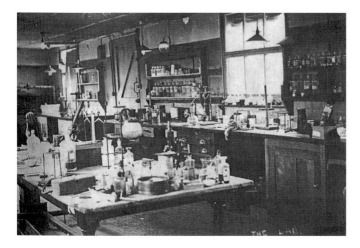

Fig. 82. Interior of the BRS laboratory at Acton

Fig. 83. The luncheon and reading room at BRS Acton

engineering and chemical laboratories, workshops and offices – was completed and ready for occupation (Figs 82 and 83)[47]. It was only then that the BRB was in a position to commence the researches on Portland cement, concrete, sand-lime bricks, jointless floors, built-up timber beams and other subjects that the Ministry of Health had requested two years' earlier[48]. Self-evidently, this was too late to be of any use in the Addison Act housing programme.

Building research and the state 1917–1921

To sum up on the origins of the BRS: state organisations for the conduct of building research were established in 1917 and 1920 in order to provide the information on non-traditional methods of construction required for the post-war housing programme. But the work of the first, the BMRC, was both curtailed and delayed as a result of the disagreement within the government over the nature of the post-war housing programme and the role that new methods would play. It was only late in 1919, when the shortage of materials and labour required for conventional construction threatened to bring the housing programme to a halt, that a more secure organisation, the BRB, was authorised; but it was not until mid-1921 that its research station at Acton was ready for occupation and in a position to commence its researches (Fig. 84). The outcome was that new methods were adopted extensively for the housing programme with neither the Ministry of Health nor the local authorities involved knowing the answers to the questions originally posed by Unwin and the Tudor Walters committee in 1917. There was a further irony. Between the end of 1918 and the middle of 1921 virtually no fresh researches into new methods were commenced by the BMRC or the BRB. This period coincided almost exactly with that in which the government was fully committed to the housing programme for which the new methods, and the research, were required.

The abandonment of the housing programme in July 1921 had serious implications for building research. The Ministry of Health had been interested in research only in order to provide information on alternatives to traditional building for the housing programme and, once the housing programme was brought to a halt, its interest in the promotion of research disappeared. This change was strikingly demonstrated by the fact that for three years from October 1921, when SB Russell left the Ministry to return to private practice, the Ministry did not consider it necessary to be represented on the BRB[49]. Since it was at the instigation of the Ministry of Health that the BRB and the BRS had been established, this inevitably raised questions about their future[50]. It was only following the return to an energetic policy of local authority housebuilding under the Wheatley Act of 1924 that the future of the BRS became assured. In October 1924, the minister of health in the first Labour government, John Wheatley, renewed the representation of the Ministry on the BRB, with Unwin being nominated as the Ministry's representative[51]. As will be shown in chapter eleven, Wheatley's successor as minister of health in the Conservative government, Neville

Fig. 84. The entire staff of BRS Acton (including professional, technical, clerical and industrial) in 1923

Chamberlain, went much further and in February 1925 charged the BRS with finding new methods of construction for municipal housebuilding that would avoid competing with private enterprise. The outcome was the construction of a much larger experimental facility at Garston in Hertfordshire, to which the BRS removed in 1925 and where the BRS (and later the BRE) has remained ever since[52].

The overall conclusion of this chapter might therefore be presented as follows. It was undoubtedly the case that, as Julian Amery, the then Minister for Housing and Construction put it in 1971, the 'creation of the BRS in 1921 was stimulated by the housing drive that followed the First World War; Lloyd George's great cry of "homes fit for heroes"'[53]. But this housing programme was itself contested. While sections of government saw it as the essential token of a new compact with a working class trained in the use of arms, other powerful interests – industry, the City, rural landowners – vigorously opposed the housing programme and were well placed to obstruct it, notably via the Treasury. The LGB, the department responsible for housing until Lloyd George replaced it with the Ministry of Health, was sceptical, if not actively hostile, as for the most part was the building department, the Office of Works. The result was that initiatives on building research were constantly undermined. The case for a building research programme was advanced repeatedly – notably by the Tudor Walters committee in the autumn of 1917; by the DSIR advisory council at the beginning of 1919; by the cabinet in November 1919; and by the DSIR advisory council in December 1919 – but on each occasion action was delayed or curtailed by Whitehall or Westminster. Although a BRB was set up in mid-1920 and the BRS opened in temporary buildings at Acton a year later, it was by no means certain how long they would survive. It was only with the emergence of a new approach to municipal housebuilding in 1924–1925, based on Neville Chamberlain's 'New Conservatism', that the long-term future of the BRS finally became assured.

Chapter eleven

Houses of paper and brown cardboard

I went down to Wembley [the British Empire Exhibition] and startled my audience by declaring that the greatest benefactor would be the man who could show us how to build homes of paper. The Birmingham press was a little disturbed and sent a man to know what I meant but on learning that I was only advocating a house cheap enough to be scrapped when it became out of date was comforted…. Of course I put it that way pour épater le bourgeois, but the idea is sound.[1]

The Building Research Station (BRS) – re-named in 1972 as the Building Research Establishment (BRE) – is generally recognised as the first and most influential state organisation in the world dedicated to building research. In outline, its early history is well known (see chapter ten). In 1917, as the British government became increasingly committed to a major housing initiative, it was pointed out by the Tudor Walters committee that this would necessitate the use of non-traditional materials and forms of construction and therefore that research would be needed to establish which were secure and satisfactory[2]. But this seemingly cogent argument was by no means uncontested; although a building materials research committee was set up in 1918, it was not until 1921 that the BRS was established – in temporary premises at Acton, west London – and even then it was by no means clear how long, as a temporary organisation with temporary staff, the BRS would survive. The 'decisive step in the growth and firm establishment' of the BRS came in 1925 when the BRS transferred from Acton to bespoke and much larger accommodation at Garston, near Watford, where it has remained to this day (Fig. 85)[3].

Fig. 85. The Building Research Station at Garston in 1938

How did this 'decisive step' come about? Although White's hitherto unpublished text gives some clues [CD], the standard published histories by RB White (1965) and FM Lea (1971) – both life-long BRS employees – give no explanation, offering an account from which those who made the key decisions – government ministers and senior civil servants – are notably absent[4]. It might be thought, *prima facie*, that the expansion of the BRS was an innovation effected by the first Labour government in 1924; and that the promotion of building research was therefore attributable to Labour's energetic minister of health, John Wheatley. The fact that the 1924 Housing Act – the 'Wheatley Act' – included measures to encourage the use of new materials and methods of construction would appear to lend credence to this presumption. But in fact the reality was almost the exact opposite. The expansion of the BRS was instigated not by John Wheatley but by Neville Chamberlain and the Conservatives. The move to Garston followed directly from a meeting called by Chamberlain (by then minister of health) in February 1925 and its purpose was to deliver what was officially termed 'the Chamberlain programme'[5]. In other words, in so far as such events are ever attributable to a single person, the expansion of the BRS is attributable to Neville Chamberlain.

Within the history of planning, Chamberlain has generally been presented in a positive light. The pioneering town planning schemes in Birmingham before and during the First World War, the unhealthy areas committee of 1919–1921 and the creation of the Greater London regional planning committee in 1928 (not to mention the appointment of the Barlow committee in 1937): according to Gordon Cherry, these made Chamberlain a figure of 'outstanding significance' in the history of twentieth-century planning[6]. Within the housing literature Chamberlain has been a more controversial figure, largely because of the political nature of housing policy after the First World War. As Clement Macintyre put it, 'underpinning the whole of Chamberlain's housing policies in the 1920s was the promotion of the role of the private builder and the restraint of the activities of the local authorities'[7]. Consequently he has been applauded by those who share this political philosophy but viewed with some disapproval by those who do not[8].

But, whether or not Chamberlain was, as AJP Taylor maintained, 'the most effective social reformer of the inter-war period'[9], there is no questioning his importance in housing and town planning policy. First appointed to the Ministry of Health – the government department responsible for housing and town planning – by Bonar Law in March 1923, thereafter he was in a position to direct housing policy in Britain for most of the next 17 years – directly as minister of health (briefly in 1923 and then from 1924 to 1929); indirectly as chancellor of the exchequer (briefly in 1923 and then from 1931 to 1937) and as prime minister (from 1937 to 1940). Thus between 1923 and 1940 there were only two quite short periods, from January to November 1923 (during the first Labour government) and from June 1929 to August 1931 (during the second) when Chamberlain was not directly or indirectly in charge of housing policy – and even then, as we will see in the case of the former, he was able to wield a sizable influence.

What gives this topic particular interest is the worldwide role that the expanded BRS was to play after its move to Garston in 1925. As the president of the Conseil International du Bâtiment, G Blachère, put it in 1971, the BRS provided a model for state-funded building research that was to be followed 'in the Commonwealth, in Europe and now all over the world'[10]. The BRS introduced a new approach by subjecting materials and construction to scientific enquiry on a systematic basis; even if it did not actually invent building science as a discipline, it was the 'BRS that really developed the concepts of building science as distinct from architecture and engineering'[11]. The BRS's *magnum opus*, *Principles of Modern Building* (published in 1938) was hailed as the first systematic attempt to treat 'building as a science-based technology' and it had an impact spanning the century, with a second edition in 1939 (reprinted seven times), a revised edition in 1959–1961 and re-workings, by DAG Reid and Steven Groák, in 1973 and 1992 respectively[12]. Scientific understanding of one of the most important new materials – concrete – was transformed by the work of FM Lea, who joined the BRS in the mid-1920s and went on to become its director some 20 years later: *The Chemistry of Cement and Concrete*, based on his work at the BRS, was published in 1935 and 70 years later, as *Lea's Chemistry of Cement and Concrete* (fourth edition, 1998) is still the standard work on the subject[13].

Housing and the 'New Conservatism'

By the time of his appointment as minister of health in 1923, Chamberlain's analysis of the housing problem was clear. Something had to be done about the housing shortage bequeathed by the First World War, if only because until it was addressed no effective steps could be taken to tackle the slums; whatever the views of some of his Conservative colleagues, 'doing nothing' was therefore not an option. The eventual answer to the housing shortage, he believed, would come from the revival of private enterprise housebuilding, but until that time temporary measures were needed to stimulate private enterprise activity. Given the existence of the rent controls inherited from the war (which Chamberlain aimed to abolish, but in stages, to minimise the political fallout) landlords could not be expected to provide the necessary demand and therefore steps were needed to encourage owner-occupiers to do so. Chamberlain summarised his views in May 1924 when his successor as minister of health, John Wheatley, had unveiled his proposals for what would become the 1924 Housing Act. Chamberlain wrote to his sister Hilda:

> I would have gone on speeding up private enterprise in building houses to sell and encouraging owner occupiership. The constant pouring of new houses into the pool by the easiest channel would have loosened the whole position.

> More and more of the cream of the working classes would have entered the new houses and the L.A. [*sic*] could have built a certain number of tenements and done something in the way of improving the slums which in time would have solved the problem.[14]

This was Chamberlain's analysis of the housing problem when, following the unceremonious eviction of Lloyd George from the premiership, Bonar Law formed his 'government of the second eleven' in October 1922. Initially Chamberlain was appointed postmaster-general but the minister of health, Sir Arthur Griffith-Boscawen, carried little weight (even with the prime minister, apparently) and when the cabinet came to consider housing in January 1923 it adopted Chamberlain's policy. This was before Chamberlain was actually appointed minister of health, in March 1923[15]. The outcome was the Housing Act of 1923, which Chamberlain brought to the statute book in July that year. The main ingredient was a new subsidy of £6 per annum (equivalent to £75 per house) for houses completed by October 1925, whether for sale or to rent. The subsidy was channelled through local authorities but they could only build themselves, rather than pass it on to private enterprise builders, with the approval of the minister of health. In addition, local authorities were given powers to subsidise the mortgage deposits and repayments of owner-occupiers. As Chamberlain later reminded the House of Commons, with the 1923 Housing Act he was not aiming to solve the housing problem as such but rather to 'put into operation machinery which would in time solve the housing problem.... namely the encouragement of private enterprise and the stimulation of the desire ... amongst large sections of the nation to be able to own their own houses'[16]:

While Chamberlain's housing bill was going through parliament, Bonar Law resigned due to ill-health and was succeeded as prime minister by the chancellor of the exchequer, Stanley Baldwin. In August 1923 Baldwin appointed Chamberlain as chancellor and two months later decided on a major change of fiscal policy – protectionism or 'Tariff Reform' – for which a mandate was promptly sought from the electorate. But the 'Protection Election' of December 1923 proved a disaster for the Conservatives who, while remaining the largest party in the House of Commons, lost their overall majority. Following defeat on a vote of confidence in January 1924, the government resigned and the King asked Ramsay MacDonald to form a (minority) government.

The advent of the first Labour government prompted a major and rapid re-think by the Conservatives of their political philosophy and image – which Chamberlain in May 1924 called 'the new spirit of conservatism' and which others have called the New Conservatism[17]. In essence this combined Baldwin's gifts for communication with Chamberlain's conviction that the espousal of social reform was the way to 'defeat socialism', ie the Labour party.

Chamberlain had long regarded social reform as the best way of defeating socialism. In 1922 he had told his sister Ida:

though of course we can't outbid the socialists I do think we should formulate something constructive without involving ourselves in heavy expenditure. My programme included Agriculture ... Women's Legislation (e.g. Children of Unmarried Parents & legislation of adoption), Poor Law Reform, Trade Union legislation (secret ballot and political levy), slums & house purchase. It's a nucleus anyway and if other people set their minds to it we ought to be able to evolve something.[18]

Chamberlain played a central role in forming the political philosophy of the New Conservatism, drawing up the statement of Aims and Principles which was accepted at the leaders' conference on 1 May 1924[19]. Published the following month as a pamphlet (and soon afterwards to provide the basis for the party's election manifesto), Chamberlain's text provided a succinct if unexciting summary of the Conservative position on foreign and domestic issues, with a strong emphasis in the latter on social reform, including unemployment relief, agriculture, housing, pensions and education[20].

For Chamberlain and the Conservatives, the advent of the Labour government raised not just the question of social reform in general but also the specific problem of responding to the proposals on housing then being laid before parliament by Labour's minister of health, John Wheatley. Wheatley's analysis of the housing problem was very different from Chamberlain's. For Wheatley the main obstacle was the shortage of skilled labour in the building trades and the answer was to get the employers and unions to agree to relax restrictions on apprenticeships, in return for the guarantee of work provided by a long-term housing programme[21]. On 6 February 1924 Wheatley set up the National House Building Council, a joint committee comprising the two sides of the industry, the National Federation of Building Trade Employers (NFBTE) and the National Federation of Building Trade Operatives (NFBTO), and in April 1924 issued its report, which recommended changes in the apprenticeship rules in return for the government undertaking a 15-year programme to build 2.5 million houses[22]. To deliver this ambitious output target and keep the rents within the means of the working class, a new and more generous subsidy would be introduced: £9 for 40 years in non-agricultural areas, equivalent to £160 per house (this at a time when the average cost of a non-parlour house was £415)[23]. But, in an adroit party political move, Wheatley also announced that, rather than abolish the subsidy provided by Chamberlain's Housing Act of the previous year, he would extend it to 1939, while at the same time removing the presumption in favour of private enterprise.

Chamberlain and new methods of construction

The unveiling of Wheatley's housing proposals in May 1924 created a problem for Chamberlain and the Conservatives. Given the claims of the New Conservatism being enunciated at the same time, they could hardly oppose such a major piece of social reform outright; but by promoting municipal housebuilding Wheatley's scheme threatened the revival of private enterprise housebuilding central to the Conservative strategy. Chamberlain wrote to his sister Ida on 18 May 1924:

> What are we to do? I haven't discussed it yet with anyone but my instinct is to let them try & fail. If we oppose we shall be accused of blocking a plan to give everybody houses in a short time out of mere partisanship or meanness & our conduct will be contrasted with Wheatley's broad mindedness in continuing my Act & even extending his own subsidy to private enterprise – which of course is mere window dressing, as it is clear that p.e. wont build to let. I incline therefore to prophesy disaster but say you are the Govt. We are not going to stop you trying to carry out your policy but we wash our hands of the results.[24]

The most likely outcome, he believed, would be a scramble for labour and materials between local authorities and private enterprise that would drive up prices and eventually bring housebuilding to a halt.

By the time Wheatley's bill reached committee stage in the Commons, Chamberlain had developed a much clearer plan, centred on the adoption of new methods of construction. He wrote to his sister Ida on 19 July:

> I went on Friday to see a model of a house which Lord Weir has very ingeniously designed for mass production. It is really a timber house with a steel plate lining on the outside and what is called 'beaver board' on the inside, this being a preparation of wood pulp. No skilled labour is required and all the materials can be provided without much extra plant. The cost will apparently be about £300 for a non-parlour house but it might be still further reduced…. I see no difficulty about the production and the model seemed to me to be satisfactory in appearance. I should think it would be rather more susceptible to heat & cold than a brick house but not much. The difficulty I see would be to get the local authorities to order in large quantities until they had thoroughly tried it out which I should think would take 18 months or two years.

Here was the answer not just to Chamberlain's parliamentary quandary but to the Conservatives' need for a policy on municipal housing that was something other than simply negative. He continued:

> But still something may come of it and I moved for an amendt [to Wheatley's bill] to provide for a reduction of contributions in case a cheap house could be produced. Wheatley refused to accept it on the ground that it wasn't necessary, which is absurd seeing the amendt couldn't do any harm if it were not necessary to use it. I think I shall move it again on report and have told Weir not to show the model to W. until after the 3rd reading…. [25]

Lord Weir was not the first inventor, nor Chamberlain the first politician, to believe that new methods of building held the key to reducing the cost of housing. The search for 'the cheap cottage' constructed of non-traditional materials went back at least 20 years, to the Cheap Cottage Exhibition held at the fledgling Letchworth Garden City in 1905 (see chapter nine)[26]. Government interest in new methods also pre-dated the First World War, for the remit given in 1912 to the departmental committee on buildings for smallholdings (of which Raymond Unwin was a member) included the investigation of 'the possibility of the reduction of cost by the use of materials and methods different from those ordinarily used' [CD][27]. During the war the shortage of clay bricks and timber compelled the government departments charged with providing housing for munition workers to employ non-traditional methods, mainly concrete blocks, most notably at the Hardwick estate at Chepstow (1916–1918) designed by William Dunn and William Curtis Green[28] (Fig. 16), and after the war the same logic led the Ministry of Health to adopt new methods for the far larger 'homes fit for heroes' campaign (see chapters one and ten). From 1919 until the axing of the programme in 1921, through the pages of its journal *Housing* and by more direct means, the Ministry pressed municipalities to adopt one of the many new systems on the market for their housing schemes. In April 1919 a standardisation and new methods of construction committee was set up to identify systems that carried the Ministry's approval; the committee was chaired by Dunn and included most of the major names – Unwin, Russell, Leonard, Scott, Ayrton, Tyrwhitt – and within a year it had approved 75 systems [CD]. Most of the systems were concrete of various sorts, mostly block (eg Crittall's Unit system which, like many others, used winget presses to produce the blocks on-site) but there were also panel systems, with fabrication either on-site (eg Duo-Slab) or off-site (eg Waller). Factory production was also a feature of the system that for a time (1919–1920) was most favoured by the Ministry, the Dorlonco system, which combined a structural steel frame with walls formed of concrete attached to metal lathes (see chapter two and Fig. 20)[29].

What distinguished Lord Weir's house from the myriad of others hitherto developed was not any inherent superiority but the overtly ideological objectives of its inventor and the charmed circles within which he moved. Weir was 'an aggressively anti-union Glasgow industrialist'[30] who regarded the stranglehold exercised by the building unions as the main cause of the country's housing problems and he saw it as his mission to devise a form of construction that bypassed the building trades altogether. He was also exceptionally well connected and moved with ease among the first division of politicians and civil servants. Thus when Bonar Law became prime minister in the autumn of 1922, he asked Weir for his ideas on housing; in February 1923 Weir had a meeting with the permanent secretary at the Ministry of Health, Sir Arthur Robinson, to discuss his plans for a standardised house; and when Chamberlain became minister of health the following month he too had a meeting with Weir, whom he asked to produce a 'demonstration house so that we may see what he is able to offer'[31]. But according to Weir's testimony in October 1924, 'on account of the Chamberlain Bill [ie the 1923 Housing Act] nothing was done with my suggestion, and I laid it on one side until the spring of this year when I was encouraged by Mr Churchill ... to again take up the matter'[32].

Even if it had a shorter life than a traditionally built house, for Chamberlain the Weir house offered a number of benefits. As he told the House of Commons in December 1924, if local authorities adopted the Weir house, they would no longer compete with private enterprise housebuilders for labour or materials; the reduction in cost would bring the houses within the means of the working class and eliminate the need for a subsidy; unemployed labour in the engineering trades would be engaged in useful work; and, by reducing the housing shortage, the time would be brought forward when it was possible to tackle the slums[33]. But the Weir house also offered another, even more important benefit that was to be aired only in private: it provided the opportunity to turn the tables on Wheatley and rouse public opinion against the trade unions. This became clear when the building union went out on strike in Glasgow in December 1924 in protest at the use of non-union labour on Glasgow Corporation's Weir houses[34]. By securing the agreement of the building unions to his ambitious housing programme, Wheatley had put the unions on the same side as those suffering from the housing problem. In contrast, by provoking the opposition of the building unions to seemingly such a 'sensible' answer to the housing shortage, the Weir house set the trade unions against the public – something that the national press could be relied on to exploit. This antagonism between unions and the electorate – particularly the women voters enfranchised in 1918, to whom Chamberlain and the Conservatives gave particular attention[35] – could only benefit the Conservatives and harm Labour. As Chamberlain told his sister in February 1925 – when, back at the Ministry of Health, he was pressing local authorities to adopt the Weir house – the Weir house was not the only type available; other new types had also been developed:

> Between ourselves the new types are good as houses, but useless for my purpose because they are quite as costly as brick if not more so.... They do not therefore excite the same hostility among the Unions who are not afraid of them. But they dare not give their real reasons for fearing Weir.... Anyway I returned to the charge when I spoke in Birmingham & I shall probably rub it in again tomorrow. My purpose of course is to work up public opinion.[36]

This, however, is to move ahead. To return to the summer of 1924 and the progress of Wheatley's bill: despite the mass of minor amendments to which it was subjected (Wheatley accepted 66 of the 76 tabled by his opponents), when the bill reached the statute book on 6 August 1924 it was 'in principle unchanged from the proposals put to cabinet six months before'[37]. In one significant respect however the bill had changed, and this directly as a result of Chamberlain's intervention. Whereas Wheatley's original bill had made no mention of new materials or methods of construction[38], as enacted it gave the minister of health substantial new powers in this regard. The new £9 subsidy could be reduced (clause 10) if a local authority refused to use new methods (an amendment introduced by Wheatley at committee stage in response to Chamberlain's amendment) and also (clause 2) where new methods were being used (Chamberlain's key objective, eventually inserted as a House of Lords amendment)[39]. Chamberlain's espousal of new methods had thereby become part of what we know as the Wheatley Act.

With the use of new methods required of local authorities by his own legislation, Wheatley had little choice but to arrange for the available systems to be assessed. At the beginning of August 1924 the government accepted a motion in the House of Lords calling for an inquiry into the various methods and materials available[40] and in September 1924 Wheatley set up a committee on new methods of house construction, chaired by Lord Moir. In characteristic Wheatley fashion this was a tripartite affair comprising representatives from the employers (WH Nicholls and AG White of the NFBTE), unions (R Coppock and G Hicks of the NFBTO) and government (including ER Forber from the Ministry of Health and Sir Frank Baines from the Office of Works). The committee met for the first time on 24 September and 10 days later went *en masse* to Glasgow to inspect the Weir house, on which it soon afterwards produced a broadly favourable report, after which it proceeded to examine concrete systems [CD][41]. In addition, towards the end of October Wheatley decided to renew the Ministry's representation on the Building Research Board (BRB), which had lapsed in 1921 at the end of the 'homes fit for heroes' campaign; Raymond Unwin, the chief architect at the Ministry and one of the leading advocates of building research since the days of the Tudor Walters committee, was the Ministry's nominee[42]. But by then the days of the Labour government were numbered. In October the government suffered a major defeat in the House of Commons and the ensuing general election, held in the wake of the notorious 'Zinoviev Letter', produced a Conservative landslide. By the second week of November Chamberlain, having declined Baldwin's offer of the Treasury[43], was back at the Ministry of Health, complete with a 25-point legislative programme[44].

Housing policy and the Building Research Station

In line with Chamberlain's new thinking, the Conservatives' manifesto for the 1924 election proclaimed the adoption of new methods as the key to tackling the housing problem. The 1923 Housing Act had shown what could be done with existing sources of labour and materials, it declared:

> Something more, however, is required if the rate of building is to be materially increased, and houses are to be produced capable of being let at a rent approaching that which can be afforded by the poorer classes. This end can

only be achieved by the employment of new materials and new methods of construction.

Accordingly a Conservative government would:

> do everything possible to foster and develop the various experiments which are now being carried out in these directions, and will not hesitate, if it be necessary, to lend financial aid to bring them to early fruition, recognising that, in this way, and in this way alone, can the provision of the housing accommodation so sorely needed be secured in a reasonable time.[45]

In contrast to the 'second eleven' cabinet he had inherited from Bonar Law in 1923, the government that Baldwin formed after the 1924 election victory was a 'cabinet of all the talents'. The ex-Liberal Winston Churchill was appointed chancellor of the exchequer – at Chamberlain's suggestion[46] – while his half-brother, Austen Chamberlain, the former party leader who had been sulking outside government since the eviction of Lloyd George, became foreign secretary. Major figures like AJ Balfour (former prime minister) and Lord Curzon (former foreign secretary) were relegated to minor roles as respectively president of the board of education and lord president of the council.

Although housing headed Chamberlain's social reform agenda, there was no immediate need for fresh legislation, thanks to the impact he had already had on the 1924 Housing Act. His intention was to retain the subsidies offered by both the 1923 Act (as amended by Wheatley) and the 1924 Act until such time as private enterprise revived and the subsidies could be phased out[47]. In terms of immediate action, the priority was to get local authorities to adopt new methods – which for Chamberlain meant the Weir house – for their housing schemes.

Chamberlain's return to the Ministry of Health (itself a token of the serious-ness with which the new government took its commitment to social reform) was widely welcomed[48]. On 15 November he wrote to Hilda:

> I have been overwhelmed with bouquets…. I have already started to plan out a four year programme. Of course the public thinks of nothing but Housing but I have told my people to make no mistake. If that had been the only problem I might not have chosen to come back….

But this did not mean housing was being ignored:

> As to housing, I have lost no time. I had a very interesting talk with Weir and have got a scheme for popularising his steel house by offering specimens to any local authority that likes to ask for them at a specially low price. I shall go and see them myself soon but from all I hear of them they seem likely to fill the bill even better than I thought.[49]

Crucial to this was the attitude of the Treasury: would the chancellor agree to this expenditure? The appointment of Churchill as chancellor had been instigated by Chamberlain and it is not hard to see why, since Churchill was keen to pursue a social reform agenda that would help justify his defection from the Liberals. As Churchill told the deputy cabinet secretary, Thomas Jones, 'I want this Gov not to fritter away its energies on all sorts of small schemes; I want them to concentrate on one or two things which will be big landmarks… Housing and Pensions'[50]. Thus when Chamberlain and Churchill met on 26 November 1924 to discuss their plans they were 'in the most complete accord'[51]. Both regarded Weir's houses as the answer; Churchill talked about building 700,000 or 800,000 houses in four years and, when told of Chamberlain's plan to offer them to local authorities at a discount, volunteered to find £50,000 for the purpose[52].

As Chamberlain noted, the problem with Weir's scheme for mass production of the steel houses was that it depended on mass orders; unlike his hero Henry Ford, Weir could not simply mass-produce the goods and wait for individual customers to come along and buy them one by one. Chamberlain's answer was to generate demonstration houses for public exhibition; once the public had seen the houses, he believed, the pressure on local authorities to go ahead and build them would be irresistible. For these demonstration houses, Chamberlain decided to offer the very substantial grant of £200 per house; if Weir delivered on his promised price of £300 per house, this would allow local authorities to acquire houses for a quarter of the normal cost. To circumvent the Fair Wages provisions contained in the Wheatley Act (which would have meant using union labour and thereby negated the *raison d'être* of Weir's scheme) the grant would be available only for houses built outside the Wheatley Act[53].

Chamberlain unveiled the scheme for demonstration Weir houses in the House of Commons on 16 December 1924 (Fig. 86). Notwithstanding the 'statesmanlike' attempts by his predecessor to expand the supply side of the industry[54], he said, the resources of the building trades were simply insufficient to meet the housing shortage and therefore there was no alternative but to look at new methods of construction:

Fig. 86. Typical Weir houses, as built by local authorities 1925–27

> Last week I paid a special visit to Glasgow in order to see for myself what was being done in this direction by Lord Weir and by others who are designing and building houses of different types, and I came back feeling very hopeful of the future.

To call it a steel house was a misnomer. It was:

> really a house constructed with a timber frame, lined with steel plates upon the outside, and upon the inside with a material which I think is composed of compressed wood pulp and asbestos but which looks like a very thick and smooth brown cardboard.

While it was common knowledge that the Weir house was the one favoured by the government, care had to be taken not to acknowledge this explicitly; the many other types available had to appear to be in contention. Accordingly Chamberlain continued:

> I am not committed to approval of this particular type to the exclusion of any other type. I think there will be probably a considerable number of houses submitted for public approval under new methods but ultimately we shall perhaps get down to two or three types.[55]

How was this process of testing and approving new systems to be carried out – and who was to undertake it? One possibility was Wheatley's committee on new methods of construction, which followed up its report on the Weir house with a second interim report on concrete systems (about which it was notably enthusiastic, particularly favouring the poured in situ method as used in The Netherlands) and a third interim report on steel and other systems (steel frame, timber frame and clay block systems as well as alternatives for roof coverings such as asbestos) [CD][56]. At first Chamberlain favoured this route and early in December 1924 strengthened the committee's member-ship by adding the contractor and former director-general of housing at the Ministry, Sir James Carmichael, as well as the London County Council's chief engineer GW Humphreys and others[57]. But the committee was dominated by the two sides of the building industry – unions and employers – and even with Chamberlain's reinforcements it lacked the scientific knowledge to give a soundly based opinion on the new methods being developed[58].

What was needed, as a number of MPs observed, was a body that could subject the various new methods and materials to scientific scrutiny[59]. The nucleus already existed in the BRS at Acton (see Fig. 84) operated by the BRB, part of the Department of Scientific and Industrial Research (DSIR).

As shown in chapter ten, the BRS had been set up in 1921 as a belated response to the call, first made by Raymond Unwin and the Tudor Walters committee in

1917, for the government to undertake research to establish which of the non-traditional methods could be used safely for the post-war housing campaign[60]. Located on a site in west London supplied by Acton Borough Council, the BRS consisted of temporary wooden sheds standing on a site the lease of which could be terminated at six months' notice either way (see Fig. 81). Since the abandonment of the 'homes fit for heroes' campaign in 1921, the BRS had occupied itself mainly with pure, rather than applied, research and with special projects such as the acoustics of the new parliament building in New Delhi. As Sir Frank Heath, the permanent secretary of the DSIR, reminded Curzon, the lord president of the Council, in February 1925:

> The present building is temporary, and the lease short, because when the Building Research organisation was first established in the Department, the Advisory Council were not at all sure whether there was sufficient ground to justify a permanent commitment to this field of work.[61]

The BRS however did have an exceptionally able and energetic director of research, RE Stradling, who had been appointed in the summer of 1924. Stradling was a young civil engineer with a PhD on the fire resistance of concrete who had been head of the department of civil engineering, architecture and building at Bradford Technical College before transferring to the BRS[62].

It was to the BRS and Stradling that Chamberlain now turned. For the BRB's bi-monthly meeting on 9 January 1925, Stradling submitted a statement on the 'increasing demand for special investigations for outside bodies and on the extent to which the Office of Works and Ministry of Health are likely to look to the Building Research Station for work of this nature'. To meet this demand he believed an additional sum of £2,500 might be required for the financial year 1925/26. But this was not all. At the meeting:

> The Director further referred to a discussion which had recently taken place between Sir Frank Heath [of the DSIR] and Mr ER Forber of the Ministry of Health on the question of the desirability of investigations being undertaken on a much larger scale than at present at the Building Research Station, and stated that a request to this effect might be received from the Ministry of Health which if approved would involve additional expenditure of a much higher order than £2,500 per annum.[63]

On 12 February, in the course of the parliamentary debate on the supplementary estimate for the grant for the Weir houses, Chamberlain announced that he was considering expanding the building research department to provide the necessary technical knowledge on new methods. The following day Heath and Stradling attended a conference with Chamberlain at the Ministry of Health 'to consider with him the extent to which the Department could assist the Ministry of Health with scientific advice in relation to the housing problem'[64]. For the meeting Stradling prepared a draft research programme. Heath reported that 'Mr Chamberlain went through the programme in great detail, and at the close of the conference decided to ask the Department officially whether they could help him along the lines discussed'[65]. The formal request followed:

> The Minister is anxious that steps should be taken which will enable a scheme
> of research, generally on the lines discussed at the conference, to be undertaken
> as soon as possible. He trusts therefore that the matter may be considered by
> your Committee with a view to arrangements being made for bringing such a
> scheme into operation at a very early date.[66]

The DSIR responded with alacrity. One week later the advisory council met to consider the request, for which it was envisaged that additional funding of about £15,000 – the same as the existing budget of the BRS – would be required, and decided to recommend it to the lord president of the council[67]. Given that the existing facilities of the BRS were 'inadequate to meet this demand', steps would have to be taken without delay to secure a site and buildings suitable for the enlarged BRS. At the same time the advisory council recommended that the formal position of the BRB should be settled by making it a permanent body[68].

Meanwhile officials at the DSIR looked into the ways in which the expanded facility could be provided. According to Heath there were four ways of expanding the facility to meet the requirements of the 'Chamberlain programme': by expanding the temporary laboratories at Acton – but this was unwise given the short lease on the site; by finding temporary accommodation nearby – but nothing was available; by building new laboratories – but this would take 18 months to two years to complete and would cost too much; or by acquiring a new site with an existing building, which could be converted much more quickly and for a much smaller sum. A suitable property at Garston, near Watford, comprising a large house and 37.5 acres (15.1 hectares) of land had been identified which could be bought for £8,000 and converted for £3000–4000; this, he said, compared to a cost of more than £30,000 for 'even the initial unit' of a purpose-built facility. Given the urgency of the matter, Heath advised Curzon that the first step would be to secure Treasury approval for the purchase and conversion of the house at Garston[69].

Curzon was by no means a friend of building research; during a previous term as lord president in 1919, he had intervened to veto the proposal for the creation of the BRB, thereby delaying the establishment of the BRS by a year[70]. This time, however, the political imperative to proceed with the government's programme was irresistible. While still opposed to the notion of open-ended building research and declining therefore to make the BRB formally permanent ('I can see no sufficient reason for giving them an unlimited extension'), on 2 March 1925 Curzon informed Heath that he agreed 'to approaching the Treasury about the Watford house'[71].

The move to Garston

The decision to proceed with the expansion and transfer of the BRS raised two immediate issues: the practical issues surrounding the conversion of the new premises and the transfer from Acton; and the political and intellectual issues involved in defining the research programme. By the end of March 1925 the freehold of the Garston property had been acquired[72] but hardly had formal Treasury sanction for the expenditure been secured than the estimated cost of the works soared: from the original figure of £3000–4000 to £10,500, which included £2680 for repairs and alterations to the existing buildings, £1515 for a new testing laboratory and £170 for a flat for Stradling in the house[73]. The Treasury complained but acquiesced; the DSIR also expressed dismay but blamed the Office of Works which, as the government's building agency, was responsible for the project and had supplied the figures.

The agreed date for the completion of the works and the transfer of the BRS to Garston was 15 September 1925 and on this basis new members of staff were appointed. But by October the works were still not finished. The Office of Works blamed the delay on the discovery that new gas and electricity mains were needed and also stated that the repairs to the mansion had proved much more extensive than anticipated. In mid-November the DSIR asked for 'definite dates by which the main building, stable block and the testing laboratory ... will be ready for occupation by the staff' (Figs 87 and 88)[74]. By December the main building and stable block were ready and the staff were able to move in before Christmas[75]; but the new testing laboratory (Fig. 89) was not finished until the end of March and so it was not until April 1926 that the transfer from Acton could be completed and the planned researches, described in the BRB's 1926 report [CD] could really get under way. On 30 April 1926 Chamberlain inspected the Garston facility and pronounced himself 'very favourably impressed' with what he saw[76].

As regards the research to be undertaken, following the meeting with Chamberlain in February 1925, Stradling drew up a detailed programme of research. This covered new and existing materials (including bricks, asbestos cement, plaster substitutes, floor and roofing materials, and products made from slag and clinker); new forms of construction (heat transmission though, and condensation on, walls; weathering etc); and the production of new specifications tailored to the needs of housing[77]. But before this could be adopted the expenditure had to be sanctioned by the Treasury (achieved on 7 April 1925)[78] and the proposed researches had to be agreed, not just with the Ministry of Health but also with the government's building department, the Office of Works.

The differences between these two departments were represented and, to some extent, exacerbated by the two personalities involved, the chief architect of the Ministry of Health, Raymond Unwin, and the director of the Office of Works, Sir Frank Baines. Unwin and Baines had long differed markedly in their views on housing and construction. In 1917–18 both had been members of the Tudor

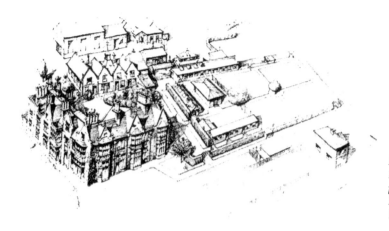

Fig. 87. The BRS at Garston, perspective by Norman Davey, 1926

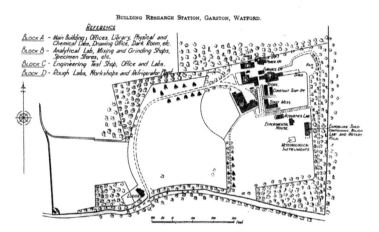

Fig. 88. The BRS at Garston, site plan, 1926

Fig. 89. The 1925 testing laboratory, exterior and interior

Walters committee but it was Unwin who had moulded (and largely written) the famous report, with Baines resigning from the committee a month before its publication; and it was Unwin who had taken the lead in promoting new methods and building research, through his membership of successively the Tudor Walters committee (1917–18), the Building Materials Research Committee (1917–1920), the standardisation and new methods of construction committee (1919–1920) and the BRB (1924–) (see chapters one and ten). In 1925 the differences between Unwin and Baines were again pronounced. Unwin's approach was pragmatic and down-to-earth, focusing on the practical requirements of delivering municipal housing, whereas Baines' view was more Olympian. In a 22-page paper on 'Research in Buildings' dated 6 March 1925, Baines argued that the piecemeal approach hitherto adopted (in which he included the Weir house) had shown itself incapable of producing a satisfactory alternative to traditional construction. What was needed was something much more fundamental, a scientific breakthrough that would deliver 'a house of new construction, which would be a much more formidable competitor in every way to the brick house'[79]. This quest for a scientific breakthrough, based on fundamental research, should be undertaken by the BRS, working of course with the appropriate government department – ie Baines' Office of Works.

Five years before, when the BRB was originally being set up, the Office of Works had been blatantly obstructive but this had profited it little; this time Baines' more nuanced approach was much more effective. Baines cultivated Stradling and under his influence Stradling re-worked the programme to give greater weight to fundamental research, moving the emphasis away from the empirical research demanded by the Ministry of Health[80]. Unwin was out of the country during April 1925 (attending the International Garden Cities and Town Planning Federation conference in New York) and on his return early in May wrote to Stradling, re-stating the Ministry of Health's position.

> While I agree as to the importance of the various proposed investigations …
> I think it should be realised that for the Ministry of Health – in this, no doubt,
> differing widely from the Office of Works – there are at the present time questions
> of such outstanding importance and urgency as to put in the background of our
> minds many matters … not having the same vital bearing on our immediate
> difficulties.[81]

The most pressing issues were the need to find alternatives to brick for wall construction and, even more, to plaster as a wall finish: these 'are the matters of most vital importance to us, and they have an urgency in time quite of a different character to that of any other matter'[82]. After these came the need for a specification for the use of breeze for concrete; for a substitute for tiles and slates for roof coverings (which would become pressing if the brick and plaster problems were solved); and for a flooring material for use on solid floors that was immune from dry rot[83].

Fig. 90. The experimental house at BRS Garston, 1926. As required by Unwin and the Ministry of Health, the steel frame allowed walls of different types to be inserted for testing – here, nine inch [22.9 centimetre] Fletton bricks set in lime mortar rather than Portland cement

To resolve these differences a meeting at the Ministry was arranged involving Stradling, Baines, Unwin and the assistant secretary at the Ministry, ER Forber. General approval was given to Stradling's programme, with the modification that the search for plaster substitutes 'should be placed first on the list' (Fig. 90)[84]. According to Unwin's notes, there was a discussion 'at some length' on the merits of fundamental versus applied research, with Baines opposing and the Ministry representatives supporting applied research. Eventually it was 'finally decided that both branches of the work could be carried on concurrently'[85].

Armed with this compromise and this programme, Stradling set about recruiting his staff. Whereas at Acton under his predecessor the scientific, professional and technical staff had numbered only eight, by the time of the move to Garston in December 1925 it had risen to 40[86]; by the end of 1926 this had increased to 60 which, with 12 office and 25 industrial staff, made a total of just under 100 (Fig. 91)[87]. By the end of 1929 the total staff would grow to 133 and by 1936 to around 200[88]. As the photograph of the scientific and technical staff outside their Garston home indicates, for the most part the recruits were young (and male) scientists, with little or no experience of building. As FM Lea recalled, 'the young staff at Garston, inspired by Dr Stradling, saw themselves as pioneers.... The ambition which stimulated them was to develop the science of building'[89].

Fig. 91. The scientific and technical staff of the BRS at Garston, c1927

Building research and social democracy

Science is often conceived as an area of human undertaking from which politics is absent. The early history of the BRS, however, indicates that, in the establishment of the science of building, politics played a major role.

During the latter stages of the First World War, in 1917, Unwin and the Tudor Walters committee had called for the government to undertake the building research required for the planned post-war housing programme. But at this stage there was powerful opposition to the housing programme and far less was done than the Tudor Walters committee had envisaged. The BRS was eventually established in 1921 but only as a temporary organisation with temporary staff.

It was only with the emergence of a new political consensus around social reform in the mid-1920s that the future of building research became secure. The key events here were the advent to power of the Labour party and, equally important, the response of the Conservatives. Municipal housing was a key Labour demand and the Housing Act of 1924 was arguably the major achievement of the first Labour government. In the changed climate brought about by the rise of Labour, the Conservatives decided to embrace social reform as a central plank of their platform. Accordingly, for electoral reasons if no other, Wheatley's subsidy had to be retained, even though it ran counter to Chamberlain's strategy for solving the housing shortage by encouraging private enterprise housebuilding.

The promotion of new methods – and, in consequence, of building research – offered the way to square the circle (Fig. 92). It allowed local authorities to carry on building without depriving private enterprise housebuilders of labour and materials. It gave the Conservatives a policy on municipal housing that

Fig. 92. London County Council, White Hart Lane estate, Tottenham, John Laing Easiform concrete cottages, 1926

was neither merely negative nor a 'me too' version of that of the Labour party. And it put the building unions and the public into opposed camps and placed the 'unreasonable' and 'obstructive' activities of the unions centre-stage. This ideological aspect of new methods made it a win-win for the Conservatives: either the unions gave way and the houses got built at lower cost, in which case the government got the credit; or (as happened with the Weir house in 1925–27) the opposition of the unions limited their use (in the event, less than 2000 Weir houses were built), in which case the trade unions could be blamed[90].

Building research undertaken by the state was, to this extent, the talisman of the new social-democratic confluence that emerged in Britain in the mid-1920s and lasted for around half a century. The Second World War re-activated and increased the impetus for social reform and for 30 years after 1945 the promotion of new methods of construction and, with it, of the work of the BRS formed a central plank in the housing policy of successive governments in the UK[91]. A similar pattern was to be seen in many other countries, particularly in Europe and the English-speaking world, which, to varying degrees, moved towards state provision of housing in this period[92]. In Britain, when that social-democratic confluence over the welfare state was ruptured by Margaret Thatcher, the rationale for building research was undermined; and it was in keeping with this changed outlook that in 1997, in one of the final acts of the Thatcher-Major government, the BRE was transferred to the private sector. That, however, is a subject for another occasion.

Endnotes and references

Abbreviations

BAL   British Architectural Library, London

BREA   Building Research Establishment Archives, Watford

FGCHM First Garden City Heritage Museum, Letchworth

HLRO   House of Lords Record Office, London

IISH   International Institute for Social History, Amsterdam

NA   National Archives, London

NCO   Nuffield College, Oxford

PP   Parliamentary Papers

PC   Private Collection

RIBA   Royal Institute of British Architects

UMDAP University of Manchester, Department of Architecture and Planning

For all books, place of publication is London except where indicated.

Introduction

1.   NA, CAB 24/111 CP 1830 (2 September 1920), quoted in SR Ward, 'Intelligence surveillance of British ex-servicemen 1918–20', *Historical Journal* 16 no 1 (1973) p179.

2.   For the USA see National Housing Association, *Housing Problems in America: Proceedings of the Seventh National Conference on Housing, Boston, November 25–27 1918* (New York, 1919); R Lubove, 'Homes and "A Few Well Placed Fruit Trees": An Object Lesson in Federal Housing', *Social Research* 27 (1960) pp469–86. The campaign for the USA to follow the British lead in housing was orchestrated by Charles Harris Whitaker, the editor of the *Journal of the American Institute of Architects*, which throughout 1918 gave detailed coverage of Hilton, Yorkship and the other US schemes.

3.   *Parliamentary Debates, Commons* 114, col 1763, 7 April 1919 (J Gilbert), quoted in M Swenarton, *Homes fit for Heroes: the politics and architecture of early state housing in Britain* (1981), p86. For Unwin and the three 'strands' of the garden city movement, see *ibid*, pp5–11.

4.   See JWR Whitehead and CMH Carr, *Twentieth-Century Suburbs: A Morphological Approach* (2001), pp80–1; M Clapson, *Suburban Century: Social Change and Urban Growth in England and the United States* (Oxford, 2003), p4; M Swenarton, 'Tudor Walters and Tudorbethan: reassessing Britain's inter-war suburbs', *Planning Perspectives* 17 (2002) p268.

5.	R Banham, *A Critic Writes: Essays by Reyner Banham* (Berkeley, 1996), pp281–91. The phrase originated with Henry-Russell Hitchcock: see H-R Hitchcock and P Johnson, *The International Style: Architecture since 1922* (New York, 1932).

6.	M Swenarton, *Homes fit for Heroes: The Politics and Architecture of Early State Housing in Britain* (1981).

7.	M Swenarton, *Artisans and Architects: The Ruskinian Tradition in Architectural Thought* (1989).

Chapter one – Home front

1.	M Bowley, *Housing and the State* 1919–44 (1945). See also P Wilding, 'The Housing and Town Planning Act, 1919: A Study in the Making of Social Policy', *Journal of Social Policy* 2 no 4 (1973) pp330–1 and CL Mowat, *Britain Between the Wars, 1918–1940* (1955; 1966).

2.	The basic facts and figures are set out in *The Official History of the Ministry of Munitions* (12 vols, 1922).

3.	J Hinton, *The First Shop Stewards Movement* (1973), pp33–6; AJP Taylor, *English History 1914–1945* (1965; Harmondsworth, 1970) pp59–60.

4.	Quoted in Taylor, *op cit*, p65.

5.	Hinton, *op cit*, pp125–127.

6.	*Official History*, vol VI, pp45, 83 and 68.

7.	NA, T132/1, Treasury to Ministry of Munitions (15 June 1915).

8.	*Official History*, vol VI, pp10–11.

9.	DT Jones, 'The Well Hall Estate' (Project Report, Open University, 1975).

10.	For the different subsidy systems see *Official History*, vol VI, pp6–12.

11.	For rent fixing policy see *ibid*, pp11–16.

12.	Figures from *ibid*, p82, and London County Council, *London Housing* (1937), pp258–67.

13.	Sir Frank Baines (1879–1933) was knighted in 1919, appointed Director of Works in 1920 and resigned from the civil service in 1927 when commissioned to design the ICI headquarters on the Embankment. Sir George Pepler (1882–1954) joined the LGB in 1914 and in 1919 was appointed technical planning chief at the Ministry of Health, specialising in regional planning. HE Farmer (1865–1933) served as resident architect at Henbury for the Ministry of Munitions before his appointment as chief architect to the director general of merchant shipbuilding. He served with the Ministry of Health 1919–1921 before returning to private practice. Gordon Allen specialised in domestic architecture and is best known for his book *The Cheap Cottage and Small House* (1912), which went into many editions. William Dunn FRIBA (1859–1934) was a founder member of the Institute of Structural Engineers and an expert on reinforced concrete. He entered government service after the war leaving the practice to W Curtis Green (1875–1960) who became one of the most successful London architects of the inter-war years (RA 1940, RIBA Gold Medal 1942). Stanley Adshead (1868–1946) held the first chair of Civic Design at Liverpool, 1909–14. In 1914 he was appointed Professor of Town Planning at University College London and went into partnership with SC Ramsey, with whom he designed important housing schemes at Kennington, Brighton, Norwich and Stepney. Unwin and Abercrombie need no introduction.

14. See R Unwin, *Nothing Gained by Overcrowding! How the Garden City Type of Development may Benefit both Owner and Occupier* (1912).

15. C Sitte (trans. GR and CC Collins), *City Planning according to Artistic Principles* (New York 1965), p61.

16. 'A Government Housing Scheme: Roe Green Village, Kingsbury', *The Builder* 114 no 3909 (4 January 1918) p7.

17. Speech by Hayes Fisher, President LGB, reported in *Municipal Journal* 27 (10 May 1918), p504; National Housing and Town Planning Council resolution, reported in *ibid*, 26 (13 April 1917), pp351–3.

18. C Addison, *Four and a Half Years: a Personal Diary from June 1914 to January 1919* (2 vols, 1934), vol 2 pp210 and 214.

19. NA, MUN 5/158 (10 February 1916).

20. R Thorne, 'The Improved Public House', *Architectural Review* 159 no 948 (February 1976) pp107–11.

21. Quoted in *Building News* 111 no 3230 (29 November 1916) pp499–500. See also the highly critical remarks of the *Town Planning Review* 6 no 2 (1915) pp147–8.

22. A striking exception being the small number of picturesque units built to Crickmer's prize-winning design from the Gidea Park cottage competition of 1911.

23. PP 1913 Cd 6708 xv, 'Report of the Departmental Committee appointed by the President of the Board of Agriculture and Fisheries to inquire into and report as to Buildings for Small Holding in England and Wales, together with Abstract of the Evidence, Appendices, and a series of Plans and Specifications', p6. Unwin was a leading member of this committee.

24. PP 1918 Cd 9191 vii, 'Report of the Committee appointed by the President of the Local Government Board and the Secretary for Scotland to consider questions of building construction in connection with the provision of dwellings for the working classes in England and Wales, and Scotland, and to report upon methods of securing economy and despatch in the provision of such dwellings', para 67. For obvious reasons, the report is normally cited by the name of its chairman, Sir John Tudor Walters, MP.

25. RL Reiss, 'Development by Cul-de-Sac', *The Architects' Journal* 66 (30 November 1927) pp705–711; M Fry, 'A Quadrangular Scheme', *ibid*, 67 (18 January 1928) pp132–140.

26. NA, DSIR 3/51, Building Materials Research Committee.

27. LGB Circular (18 March 1918), reprinted in *Municipal Journal* 26 (22 March 1918) pp317 and 319.

28. RIBA, *Housing of the Working Classes in England and Wales: Cottage Designs* (1918).

29. *Building News* 115 no 3334 (27 November 1918) p352.

30. 'Housing for Munitions Workers and other Buildings at Gretna and Eastriggs', *Building News* 115 no 3326 (2 October 1918) pp222, 224–5 and 231–2; *ibid*, 115 no 3328 (16 October 1918) pp256, 262–3 and 265; *ibid*, 115 no 3329 (23 October 1918) p272; *ibid*, 115 no 3332 (13 November 1918) pp324–5. 'The Government Housing Scheme carried out at Mancot, near Chester', *The Architects' and Builders' Journal* 48 no 1251(25 December 1918), Special Supplement, pp1–32. The first proper publication of Gretna was in the Building News 115 no 3326 (2 October 1918). The scheme

had been described earlier in the USA without illustrations (by order of the censor) and, under the pseudonym 'Moortown' had been the subject of a series of enthusiastic articles in *The Times* by Sir Arthur Conan Doyle.

31.   NA, MUN 5/158.

32.   NA, RECO 1/631, WH70. Mrs Lloyd is reputed to have assisted Unwin in the preparation of *Town Planning in Practice* (1909). Her husband, Dr T Alwyn Lloyd, worked under Unwin at Hampstead before becoming architect to the Welsh Town Planning and Housing Trust.

33.   *Building News* 115 no 3334 (27 November 1918) p352.

34.   *The Architect* 101 no 2623 (28 March 1919) p224.

35.   G Allen, *The Cheap Cottage and Small House* (sixth ed., 1919), p70.

36.   'Dormanstown: An Industrial Village', *The Architects' Journal* 49 no 1273 (28 May 1919) p371.

37.   Cd 9191, *op cit*, paras 144 and 146.

38.   *ibid*, para 149.

39.   NA, CAB 24/72, GT 6552 (23 December 1918).

40.   PB Johnson, *Land Fit for Heroes: The Planning of British Reconstruction 1916–1919* (Chicago, 1968), pp338–431.

41.   NA, MH 78/64.

42.   Information from *The Architectural Who's Who* (1923).

43.   PP 1920 Cmd 917 xvii, 'First Annual Report of the Ministry of Health,1919–20, Part II, Housing and Town Planning', pp8 and 9.

44.   HA Saul and CF Simmons, both of whom worked on the Gretna project, obtained middle-rank posts. RS Bowers, of the Gretna team, went into partnership with EG Culpin, their firm becoming one of the leading practices designing public housing schemes. Theodore Fyfe, site architect at Mancot, became first head of the Department of Architecture at Cambridge in 1922. Professor Adshead wrote and lectured frequently on his scheme for Dormanstown, becoming one of the leading theoreticians of public housing design in the early 1920s. He was widely employed as a site planning consultant.

45.   EG Culpin,'The Remarkable Application of Town Planning Principles to the Wartime Necessities of England', *Journal of the American Institute of Architects* 5 no 4 (April 1917) pp157–9.

46.   *The Builder* 116 no 3961 (23 January 1919) p25.

47.   CR Ashbee, *A Book of Cottages and Little Houses* (1906), pp111–3. See also R Unwin, *Cottage Plans and Common Sense* (Fabian Tract 109, 1902) and R Unwin, 'Cottage Building in Garden City', *The Garden City* ns 1 no 5 (June 1906) pp107–11.

48.   S Adshead, 'The Standard Cottage', *Town Planning Review* 6 no 4 (1916) pp244–9. See also L Budden, 'The Standardisation of Elements of Design in Domestic Architecture', *ibid*, pp238–43.

Chapter two – Neo-Georgian *maison-type*

1. Dormanstown was published in *The Architect* 101 no 2628 (2 May 1919) pp295–6, 300–1 and 304–5; *The Architects' Journal* 49 no 1273 (28 May 1919) p372; *ibid*, 50 no 1304 (31 December 1919) pp812–4. See also CH James and FR Yerbury, *Small Houses for the Community* (1924). A copy of the consultants' original report is to be found in the library of the Bartlett at University College London. Since this essay was written, our knowledge of the Liverpool school has been amplified by M Wright, *Lord Leverhulme's unknown venture: the Lever Chair and the beginnings of town and regional planning, 1908–48* (1982).

2. R Unwin, *Town Planning in Practice* (1909).

3. For technical descriptions see Ministry of Works Postwar Building Study No 1, *House Construction* (1943), pp74–6 and RB White, *Prefabrication: A History of its Development in Great Britain* (1965), pp60–1.

4. See NA HLG 49/11, report from regional surveyor on Dorman Long houses, 17 May 1920; letter from Ministry of Health to Dorman Long, 21 September 1920; letter from Dorman Long to Ministry of Health, 23 January 1925. These technical failures first appeared in 1920, at a time when the Ministry was pressing local authorities to adopt the Dorman Long system for their housing schemes: see M Swenarton, *Homes fit for Heroes. The Policy and Design of the State Housing Programme of 1919* (University of London PhD, 1979), chapter 11.

5. M Bowley, *The British Building Industry: four studies in response and resistance to change* (Cambridge, 1966). Bowley, however, does not acknowledge the contribution of Adshead to the design of the system, which she regards simply as a manufacturer's invention. Indeed, one of the themes of this important book is that architects played virtually no part in the development of new building methods. Perhaps more detailed research into the genesis of the early 'systems' will further shake this view.

6. Stanley Adshead (1868–1946) served his articles in Manchester and London (where he worked for George Sherrin, Guy Dawber and Sir Ernest George) before setting up his own practice just before the turn of the century. Between 1911 and 1935 he was in partnership with Stanley Ramsey with whom he built the Kennington estate, Dormanstown and major postwar housing projects in Brighton, Norwich, Dover, Totnes, Newburn and Stepney. The firm also collaborated with Abercrombie on the Middlesbrough and Chesterfield plans. However, it was Adshead's taste and his skill as a draughtsman that most impressed his contemporaries. Sir Charles Reilly attributes these qualities to the good influence of his father, a painter, and records that one RA Exhibition contained no fewer than 20 of his perspectives – 'the Cyril Farey of his age'.

7. Stanley Ramsey (1883–1969) was trained at the part-time school of architecture at King's College, London (where Reilly taught before going to Liverpool) and was articled to C Stanley Peach, whose partner he became. In 1911 he went into partnership with Adshead. He was general editor of Benn's Masters of Architecture series, to which he contributed the volume on *Inigo Jones* (1924). With JDM Harvey (who took the photographs) he published *Small Houses of the Late Georgian Period, 1750–1820* (2 vols, 1919–23). He served as Vice-President of the RIBA in 1944–45.

8. *The Builder* 106 no 3707 (20 February 1914) p226; *The Architects' and Builders' Journal* 39 no 1000 (4 March 1914) pp151–4. The Architectural Association visited the scheme: see Architectural Association Journal 29 no 326 (April 1914) pp248–9.

G Stamp, 'London 1900', Architectural Design 48 nos 5–6 (1973), pp365–6, suggests that it was Ramsey's previous study of late-Georgian and Regency buildings which so strongly and favourably influenced the Kennington scheme.

9. Sir Patrick Abercrombie (1879–1957) was educated at Uppingham and articled in Manchester and in the Liverpool office of Sir Arnold Thornely. Reilly appointed him studio assistant in the School of Architecture and, in 1910, he was made research fellow in the new Department of Civic Design. He joined the RIBA – his first formal qualification – after his appointment as Professor of Civic Design at the age of 36. Between his Dublin Plan of 1913 (a competition victory) and his London plans of the mid-1940s, he took a leading role in schemes for Sheffield, East Kent, Doncaster, South Teesside, Deeside, East Suffolk, Bristol and Bath, Cumberland, Stratford, Plymouth, South West Lancashire and the Thames Valley. Numerous other commissions at home and abroad followed his County of London Plan (1943) and the Greater London Plan (1944). Knighted in 1945, he received the RIBA Gold Medal in 1946 and the AIA Gold Medal in 1949.

10. Sir Charles Reilly (1874–1948) was educated at Merchant Taylors' School and Queen's College, Cambridge, where he took a first in engineering in 1896. For a year he worked in his father's architectural office before being articled to John Belcher, attending the AA night school as a student and, during the day, teaching in the part-time school of architecture at King's College, London. In 1904, aged 30, he was appointed to the Roscoe chair at Liverpool, which he held until his retirement due to ill health in 1933. His RIBA Gold Medal (1943) and knighthood (1944) were awarded for contributions to education. Reilly was made OBE for work during the First World War, when he served as senior inspector for the ammunition division of the Ministry of Munitions. War service of this nature may well reflect his interest in national efficiency (see below). His autobiography *Scaffolding in the Sky* (1938), an engaging if not always scrupulously accurate work, gives an excellent insight into his personality and beliefs. See also S Turner and A Allen, 'The Papers of Sir Charles Reilly: a recent accession to the university archive', *University of Liverpool Recorder* (October 1979) pp159–167.

11. CH Reilly, *Scaffolding in the Sky* (1938), p122.

12. *ibid*, p126.

13. In a letter to Reilly, *ibid*, p121. See also R Macleod, *Style and Society: architectural ideology in Britain 1835–1914* (1971), pp97–8.

14. L Budden, 'The Standardisation of Elements of Design in Domestic Architecture', *Town Planning Review* 6 no 4 (1916) p239. Lionel Budden (1887–1956) was trained at the University of Liverpool where he took a first in 1909. After further study in Athens and Berlin he joined the staff as Assistant Lecturer in 1911 and remained in the school until his retirement in 1952. Unable to serve in the First World War because of poor health, he directed the school in the absence of Reilly (1915–19) and, in 1933, succeeded Reilly as Professor. Budden's classicism found an outlet in the design of numerous war memorials during the 1920s and, in collaboration with Reilly and Marshall, he built the extension to the Liverpool School of Architecture (1932) in the 'stripped classical' style which was to become the school's contribution to the English modern movement of the 1930s.

15. For the housing congress see *Municipal Journal* 25 (24 March 1916) p319 and *Housing Journal* 105 (May 1916) p5.

16. S Adshead, 'The Standard Cottage', *Town Planning Review* 6 no 4 (1916), pp244–9;

S Ramsey, 'The Small House of a Hundred Years Ago', *ibid*, 6 no 4 (1916), pp222–2; L Budden, *op cit*, pp238–243.

17. C Reilly, 'The City of the Future', *Town Planning Review* 1 no 3 (1910) pp191–7; Abercrombie, 'The Square House', *ibid*, 4 no 1 (1913) pp35–43.

18. Budden, *op cit*, p238. For the Purist theory of the *objet type*, see R Banham, *Theory and Design in the First Machine Age* (1960), chapter 15. Examples of typical declarations by Muthesius, Corbusier, Stam and others are to be found in T and C Benton with D Sharp, *Form and Function: A Source Book for the History of Architecture and Design 1890–1939* (1975), whose excellent translations are often more coherent than the original.

19. Adshead, *op cit*, pp246 and 248.

20. Unwin, *Cottage Plans and Common Sense* (Fabian Tract 109, 1902); PP 1918 Cd 9191 vii, 'Report of the Committee appointed by the President of the Local Government Board and the Secretary for Scotland to consider questions of building construction in connection with the provision of dwellings for the working classes in England and Wales, and Scotland, and to report upon methods of securing economy and despatch in the provision of such dwellings' (the Tudor Walters Report), paras 158 and 349.

21. HF Heath and AL Hetherington, *Industrial Research and Development in the UK: A Survey* (1946), pp333–6.

22. *Parliamentary Debates, Commons* 144, col 1791, 7 April 1919 (N Billing).

23. *Nineteenth Century* (May 1901) p843, quoted by GR Searle, *The Quest for National Efficiency: A Study in British Politics ad Political Thought 1899–1914* (Oxford, 1971), p52.

24. The classic legal exposition of the 'growth of collectivism' is that of AV Dicey, *Lectures of the Relation Between Law and Public Opinion in England during the Nineteenth Century* (1905) which identifies its appearance rather earlier than the turn of the nineteenth and twentieth centuries.

25. Unwin, *op cit*, p375.

26. Reilly, *op cit*, p193.

27. Searle, *op cit*.

28. L Urwick and EFL Breck, *The Making of Scientific Management* (3 vols, 1946–48), vol 1 pp32–5 and vol 2 pp88–107.

29. CK Hobson, 'Economic Mobilisation for War', *Sociological Review* 8 no 3 (July 1915) p156.

30. CA Elwood, 'The Social Problem and the Present War', *Sociological Review* 8 no 1 (January 1915) p13.

31. WH Dawson (ed), *After War Problems* (1917), p 7, quoted in A Marwick, *The Deluge: British society and the First World War* (1965), p135.

32. *ibid*, p176.

33. *FBI: What It is and What it Does*, quoted by Marwick, *op cit*, p254.

34. Adshead, *op cit*, pp244–5.

35. Budden, *op cit*, p239.

36. CH Reilly, *Representative British Architects of the Present Day* (1931), p61. See also Macleod, *op cit*, pp97–8.

37. MG Hawtree, *The Origin of the Modern Town Planner* (University of Liverpool PhD, 1975), p185.

38. R Blomfield, *The Mistress Art* (1908), pp3 and 7.

39. Budden, *op cit*, p241.

40. Adshead, *op cit*, p248.

41. Budden, *op cit*, p243.

42. *The Architects' Journal* 50 (1919), p812.

Chapter three – An insurance against revolution

1. *Public General Acts* 9 & 10 Geo 5, ch 35, Housing, Town Planning etc Act, 1919, ss1, 2 and 7. M Bowley, *Housing and the State, 1919–44* (1945), p15. HW Richardson and DH Aldcroft, *Building in the British Economy between the Wars* (1968), p164.

2. Calculated from: PP 1916 Cd 8196 xii, '44th Annual Report of the Local Government Board, 1914–15, Part II, Housing and Town Planning', pp 20–21; BR Mitchell and P Deane, *Abstract of British Historical Statistics* (Cambridge, 1962), pp236–7. The principal statute was the 1890 Housing Act: see Public General Acts, 53 & 54 Vic ch 70, Housing of the Working Classes Act, 1890.

3. Bowley, *op cit*, p271.

4. Local Government Board, *Manual on the Preparation of State-Aided Housing Schemes* (1919), p4. For prewar local authority housing, see M Swenarton, *Homes fit for Heroes: the Politics and Architecture of Early State Housing in Britain* (1981), pp34–44.

5. *Manual on State-Aided Housing Schemes*, pp4–9 and appendices I and IV; Swenarton, *op cit*, pp11–26, 141–61 and 187–8.

6. *Parliamentary Debates, Commons* 144, cols. 1483–1485, 14 July 1921 (Sir A Mond).

7. See Ministry of Health, *Housing* ii p177, p207 and p229 (20 December 1920, 17 January and 14 February 1921); NA, HLG 31/2, 'Memorandum to housing commissioners' no 136 (25 June 1921); *Municipal Journal* 30 pp355, 617, 684 and 734 (13 May, 19 August, 9 September and 7 October 1921).

8. Bowley, *op cit*, pp2–35. See also Richardson and Aldcroft, *op cit*, pp164 and 169.

9. P Abrams, 'The failure of social reform, 1918–20', *Past and Present* 24 (April 1963) p59; S Marriner, 'Sir Alfred Mond's Octopus: a Nationalised House-Building Business', *Business History* 21 no 1 (1979) pp26–7.

10. BB Gilbert, *British Social Policy, 1914–39* (1970), pp137–61.

11. This applies also to the other published studies of the housing programme: PB Johnson, *Land Fit for Heroes: the Planning of British Reconstruction, 1916–19* (Chicago, 1968) and P Wilding, 'The Housing and Town Planning Act 1919: a study in the making of social policy', *Journal of Social Policy* 2 no 4 (1973) pp317–34.

12. London County Council, Housing of the Working Classes Committee, 14 May 1919. See also NA, RECO 1/642, memorandum on housing in England and Wales by BS Rowntree; and Bowley, *op cit*, p12.

13. NA, RECO 1/645, statement by J Carmichael.

14. PP 1918 Cd 9087 xxvi, 'Housing in England and Wales. Memorandum by the Advisory Housing Panel of the Ministry of Reconstruction on the Emergency Problem', p4.

15. See *Municipal Journal* 26 pp495–6 (25 May 1917); *Housing Journal* 110 pp1–3 and 5 (August 1917).

16. NA, CAB 24/42 GT 3617, Housing of the working classes after the war: memorandum by the President of the Local Government Board (February 1918).

17. C Addison, *Four and a Half Years: a Personal Diary from June 1914 to January 1919* (2 vols, 1934), vol 2 p414.

18. NA, CAB 24/44 GT 3803, Memorandum by the Minister of Reconstruction.

19. *The Housing of the Working Classes Acts, 1890 to 1909, Memorandum with respect to the Provision and Arrangement of Houses for the Working Classes* (1913) and *Memorandum for the use of Local Authorities with respect to the Provision and Arrangement of Houses for the Working Classes* (1917). For the cottage competition, see NA, T 128/2, letter of 29 August 1917, and NA, RECO 1/624, letter from N Kershaw of 27 June 1918; also RIBA, *Housing of the Working Classes in England and Wales. Cottage Competitions. Conditions* (1917).

20. Minute of Rowntree to Addison, 14 September 1917, quoted in Johnson, *op cit*, p90.

21. NA, RECO 1/624, 'LGB observations on interim report of the women's sub-committee' (27 August 1918). For a discussion of one of the major criteria of housing standards – the bathroom – see M Swenarton, 'Having a bath: English domestic bathrooms, c1890–1940', in Design Council, *Leisure in the Twentieth Century: History of Design* (1977), pp92–9.

22. NA, RECO 1/624, note by Secretary of the Women's Housing Sub-committee (30 August 1918). Compare the original report, in RECO 1/624, with the published version, PP 1918 Cd 9166 x, 'Ministry of Reconstruction Advisory Council: Women's Housing Sub-committee. 1st Interim Report'.

23. See J Hinton, *The First Shop Stewards' Movement* (1973), pp196–213, and CJ Wrigley, 'Lloyd George and the labour movement' (University of London PhD, 1974) pp295–305.

24. PP 1917–18 Cd 8696 xv, 'Summary of the Reports of the Commission of Enquiry into Industrial Unrest by the Rt. Hon. GN Barnes, MP', pp6–7. See also PP 1917–18 Cd 8669 xv, 'Report of the Commissioners for Scotland', p4.

25. NA, CAB 23/3 WC 190 (19 July 1917).

26. NA, CAB 23/3 WC 194 (24 July 1917).

27. *Municipal Journal* 26 p739 (3 August 1917).

28. *ibid*, p737. For a fuller discussion, see Swenarton, *op cit* (1981), pp70–7.

29. NA, LAB 2/555/F (DR) 103, Ministry of Labour memorandum (22 February 1919).

30. HLRO, Lloyd George Papers, F/30/3/13, Bonar Law to Lloyd George (30 January 1919), quoted in Wrigley, *op cit*, p383.

31. NA, CAB 23/9 WC 523 (31 January 1919).

32. *ibid*, WC 527 (6 February 1919).

33. HLRO, Lloyd George Papers, F/30/3/32, Lloyd George to Bonar Law (20 March 1919), quoted in Wrigley, *op cit*, pp413–14.

34. NA, CAB 23/9 WC 520 (28 January 1919). See also CAB 23/9 WC 519 (27 January 1919).

35. NA, CAB 24/111 CP 1830 (2 September 1920), quoted in SR Ward, 'Intelligence surveillance of British ex-servicemen 1918–20', *Historical Journal* 16 no 1 (March 1973), p179. But it seems that labour was almost as frightened of the ex-servicemen as was the government: see D Englander and J Osborne, 'Jack, Tommy and Henry Dubb: the armed forces and the working class', *Historical Journal* 21 no 3 (September 1978) pp593–621.

36. HLRO, Lloyd George Papers, F/30/3/32, Lloyd George to Bonar Law (20 March 1919), quoted in Wrigley, *op cit*, pp413–4.

37. NA, CAB 23/9 WC 539 (3 March 1919).

38. *The Times* (13 November 1918), quoted in Gilbert, *op cit*, p19.

39. *The Times* (2 April 1919), in NA, HLG 29/117.

40. *Parliamentary Debates, Commons* 114, col. 1956, 8 April 1919 (W Astor).

41. NA CAB 23/9 WC 539 (3 March 1919).

42. *ibid*, WC 541 (4 March 1919).

43. *Parliamentary Debates, Commons* 114, cols. 1740 and 1743, 7 April 1919 (Sir D Maclean).

44. *Public General Acts* 9 & 10 Geo 5 ch 35, Housing, Town Planning etc Act, 1919.

45. *Parliamentary Debates, Commons* 114, col. 1762, 7 April 1919 (JD Gilbert). See also *ibid*, cols. 1786 (Billing), 1810 (Bethell) and 1773 (Pretyman).

46. The *Manual on State-Aided Housing Schemes* of April 1919 summarised the Tudor Walters Report of the previous November: PP 1918 Cd 9191 vii, 'Report of the Committee appointed by the president of the Local Government Board and the Secretary for Scotland to consider questions of building construction in connection with the provision of dwellings for the working classes in England and Wales, and Scotland, and report upon methods of securing economy and despatch in the provision of such dwellings' (the Tudor Walters Report).

47. A Sayle, *The Houses of the Workers* (1924), p140. For detailed substantiation, see Swenarton, *op cit* (1981), pp141–61, 169–72 and 180–2.

48. PP 1918 Cd 9182 vii, 'First Interim Report of the Committee on Currency and Foreign Exchanges after the War', p6. See also S Howson, 'The origins of dear money, 1919–20', *Economic History Review* second series 27 no 1(February 1974), pp88–107.

49. For these difficulties and government attempts to circumvent them, see Swenarton, *op cit* (1981), pp112–29; also chapter ten.

50. NA, HLG 68/29, minute by AW Robinson (30 June 1920). See also NA, CAB 24/106, memorandum by A Chamberlain (20 May 1920).

51. PRO, CAB 27/89, Cabinet Housing Committee, 1st conclusions (17 June 1920).

52. Howson, pp88–90; CL Mowat, *Britain between the Wars, 1918–40* (1955; 1966), p126.

53. NA, CAB 24/115 CP 2145, 'Second interim report of the cabinet committee on unemployment' (25 November 1920); P Rowland, *Lloyd George* (1975), pp529–30.

54. NA, CAB 23/38, conference 70 (15 December 1920).

55. Ward, *op cit*, p187.

56. NA, CAB 23/22 c.49 (20) (17 August 1920). See also GDH Cole, *A Short History of the British Working Class Movement, 1789–1947* (3rd edn., 1948, 1960), pp380–403, and RP Arnot, *The Miners: a History of the Miners' Federation of Great Britain* (3 vols, 1949–61), vol 2 pp226–338.

57. NA, CAB 23/25 c.23 (21) (18 April 1921). See also K Middlemas (ed), *Thomas Jones Whitehall Diary* (3 vols, 1969), vol 1 p158.

58. NA, CAB 27/72 FC 52, note by A Chamberlain (20 November 1920). Also CAB 27/71, Finance Committee, 28th conclusions (29 November 1920).

59. NA, CAB 27/72 FC 66, memorandum by C Addison (25 January 1921).

60. NA, HLG 68/29, 'Memorandum to Housing Commissioners' no. 85a (22 February 1921).

61. NA, HLG 68/29, letters from Chamberlain to Addison (9 March 1921) and from Addison to Chamberlain (11 March 1921).

62. Rowland, *op cit*, pp532–4.

63. HLRO, Lloyd George Papers, F/7/4/6, quoted in Rowland, *op cit*, p538.

64. NA, CAB 23/26 c.55 (21) (29 June 1921).

65. NA, CAB 27/71, Finance Committee, 35th conclusions (30 June 1921). The conclusions were altered after the meeting by the agreement of the Chancellor and the Minister of Health: for the original conclusions, see T 161/132, letter of Sir A Mond (2 July 1921). Unfortunately the date of this meeting is erroneously given as 30 January 1921 in Gilbert, *op cit*, p152, which has misled those who have drawn on his account.

66. *Parliamentary Debates, Commons* 144, cols. 1483–1485, 14 July 1921 (Sir A Mond). For Addison's attempts to resist this proposal, see NA, CAB 24/126 CP 3108, memorandum by Addison (4 July 1921); CAB 24/126 CP 3111, memorandum by Mond (7 July 1921); CAB 23/26 c58 (21) (11 July 1921); CAB 23/39 conference 105 (12 July 1921); CAB 23/39 conferences 106 and 107 (13 July 1921).

67. See Swenarton, *op cit* (1981), pp151–64.

68. See note 7, above.

69. *Municipal Journal* 31 (7 July 1922) p487.

70. NA CAB 24/84 GT 7790, report on revolutionary organizations (29 July 1919), quoted in Ward, *op cit*, p187.

71. Bowley, *op cit*, pp2–14; Richardson and Aldcroft, *op cit*, p164.

72. Wilding, *op cit*, p332; Marriner, *op cit*, pp26–7.

73. NA, CAB 23/9 WC 539 (3 March 1919) and WC 541 (4 March 1919); Howson, *op cit*, p92. Since the publication of this essay, this interpretation has been challenged by Martin Daunton (who had not studied the relevant Cabinet papers) but Daunton's view has in turn been challenged by Murray Fraser (who had). See M Daunton,

House and Home in the Victorian City: Working-Class Housing 1850–1914 (1983), p298 and M Fraser, *John Bull's Other Homes: State Housing and British Policy in Ireland, 1883–1922* (Liverpool, 1996), p299. For a broader overview, see A Ravetz, *Council Housing and Culture: The History of a Social Experiment* (2001).

74.   NA, CAB 24/126, memorandum by Addison (4 July 1921); HLG 68/29, minute on housing (8 December 1920); Bowley, *op cit*, pp11–12.

75.   NA, CAB 24/126, CP 3133, amended draft statement of the Minister of Health; Parliamentary Debates, Commons 144, cols. 1483–1485, 14 July 1921 (Sir A Mond).

76.   Bowley, *op cit*, p26; Richardson and Aldcroft, *op cit*, p169; J Burnett, *A Social History of Housing, 1815–1970* (Newton Abbot, 1978), p222.

77.   PP 1921 Cmd 1447 xiii, 'Report of the Departmental Committee on the High Cost of building Working Class Dwellings', p44. Average tender prices rose from an estimated £600 in 1918 to an actual £1200 or more in 1920: see Swenarton, *op cit* (1981), p122.

Chapter four – Rationality and rationalism

1.   N Pevsner, *Pioneers of the Modern Movement* (1936), p168.

2.   N Pevsner, *An Outline of European Architecture* (Harmondsworth, 1943, 1968), p394.

3.   S Giedion, *Space Time and Architecture* (Cambridge, Mass., 1941); HR Hitchcock, *Architecture: Nineteenth and Twentieth Centuries* (Harmondsworth, 1958).

4.   R Banham, *Theory and Design in the First Machine Age* (1960), p47.

5.   C Jencks, *Modern Movements in Architecture* (Harmondsworth, 1973); K Frampton, *Modern architecture: a critical history* (1980), p47. The picture has not changed with more recent English-language histories: for example in A Colquhoun, *Modern Architecture* (Oxford/New York, 2002) the garden city movement is mentioned several times but only in a non-UK context – Germany, France, Finland and the USA.

6.   M Tafuri, 'L'Architecture dans le Boudoir: the language of criticism and the criticism of language', *Oppositions* 3 (May 1974) p56. Previously Leonardo Benevolo in History of Modern Architecture (2 vols, Italian edition 1960, English translation 1971) had discussed Ebenezer Howard and the 'widespread influence' of the garden city movement, but he presented this as falling *outside* the mainstream of modern architectural development.

7.   M Tafuri and F Dal Co, *Modern Architecture* (1980). See notes 16 and 22 below.

8.   See W Creese, *The Search for Environment. The Garden City: Before and After* (New Haven, 1966) and, for more recent publications, see endnote 1 to chapter seven.

9.   The main writings of Unwin on which this section is based are *Cottage Plans and Common Sense* (Fabian Tract 109, 1902); *Town Planning in Practice* (1909); 'Town Planning at Hampstead', *Garden Cities and Town Planning* ns 1 no 1 (February 1911) pp6–12 and ns 1 no 4 (May 1911) pp82–5; *The Town Extension Plan* (Warburton lecture, University of Manchester 1912); *Nothing Gained by Overcrowding! How the Garden City Type of Development may Benefit both Owner and Occupier* (1912). Also the Report of the Tudor Walters Committee (for which Unwin was largely responsible): PP 1918 Cd 9191 vii, 'Report of the Committee appointed by the President of the Local Government Board and the Secretary for Scotland to consider questions of building construction

in connection with the provision of dwellings for the working classes in England and Wales, and Scotland, and report upon methods of securing economy and despatch in the provision of such dwellings' (the Tudor Walters Report).

10. *Cottage Plans and Common Sense*, p3.

11. *The Town Extension Plan*.

12. 'Town Planning at Hampstead' (February 1911) p6.

13. *Town Planning in Practice*, p11.

14. Cd 9191, para 152.

15. See H Allen Brooks, 'Jeanneret and Sitte: Le Corbusier's Earliest Ideas on Urban Design', in H Searing (ed), *In Search of Modern Architecture: A Tribute to Henry-Russell Hitchcock* (Cambridge, Mass., 1982), pp278–97. For Sitte, see GR and CC Collins, *Camillo Sitte and the Birth of Modern City Planning* (New York, 1965) which remains the outstanding study of city planning in this period – especially in its revised edition: GR Collins and C Crasemann Collins, *Camillo Sitte: The Birth of Modern City Planning* (New York, 1986).

16. *Town Planning Review* 7 nos 3–4 (March 1918) pp251–2 and plate 63. However, according to Tafuri and Dal Co (*Modern Architecture*, p133), 'in the project for workers' housing at Saint-Nicholas d'Aliermont in France, Le Corbusier finally took the path that led him to leave completely his first urbanistic conceptions that were still influenced by the garden city idea'.

17. 'Le rôle et les méthodes de l'office public' (1919), in H Sellier, *La Crise du Logement et l'Intervention Publique en Matière d'Habitation Populaire dans l'Agglomération Parisienne* (Paris, 1921), pp 253–93. See also J Read, 'The Garden City and the Growth of Paris', *Architectural Review* 163 no 976 (June 1978) pp345–52.

18. See the layout plan in Sellier, *op cit*, p683.

19. P Wolf, *Wohnung und Siedlung* (1926). On the Charlottenburg school, see R Wiedenhoeft, 'Workers Housing as Social Politics', VIA 4 (1980) p122.

20. For May, see N Bullock, 'Housing in Frankfurt 1925 to 1931 and the new *Wohnkultur*', *Architectural Review* 163 no 976 (June 1978) pp335–42; G Uhlig, 'Town Planning in the Weimar Republic', *Architectural Association Quarterly* 11 no 1 (1979) pp24–38; J Castex, J-C Depaule and P Panerai, *Formes urbaines: de l'îlot à la barre* (1977), chapters 4 and 7. This last has since been translated into English: P Panerai, J Castex and J-C Depaule, *Urban forms: death and life of the urban block* (Oxford, 2004).

21. *Das neue Frankfurt* 4 nos 4–5 (April–May 1930) p77 et seq; Castex et al, *op cit* (1977), pp119–23.

22. Congrès Internationaux d'Architecture Moderne/Internationale Kongresse für Neues Bauen (CIAM), *Rationelle Bebauungsweisen* (Frankfurt, 1931). Tafuri and Dal Co (*Modern Architecture*, p246) confuse this 1930 Brussels meeting with the Mediterranean meeting of 1933.

23. W Gropius, 'Flach-, Mittel- oder Hochbau?', in *Rationelle Bebauungsweisen*, pp26–47; translated as 'Houses, Walk-Ups or High-rise Apartment Blocks' in W Gropius, *The Scope of Total Architecture* (1955; 1956), pp117–30.

24. *Das neue Frankfurt* 4 nos 2–3 (February–March 1930) pp55–8. Also Castex et al, *op cit* (1977), pp123–9.

25. Cd 9191, para 58.

26. See *Rationelle Bebauungsweisen*, plan no 36; O Haesler, *Mein Lebenswerk als Architekt* (Berlin, 1957), pp3–40; B Miller Lane, *Architecture and Politics in Germany 1918–1945* (Cambridge, Mass., 1968), pp63, 90 and 245.

27. C Bauer, *Modern Housing* (Boston, 1934), p181.

28. Miller Lane, *op cit*, p245.

29. Bauer, *op cit*, p181.

30. For Radburn, see Clarence Stein, *Towards New Towns for America* (Cambridge, Mass., 1951; second edition, Liverpool, 1958).

31. Bullock, *op cit*, pp337–8.

32. Internationale Verbandes für Wohnungsweisen (International Housing Association), *Der Bau von Kleinwohnungen mit Tragbaren Mieten (The Building of Small Dwellings with Reasonable Rents)* by F Schuster (International Housing Congress, Berlin 1931), p8.

33. Gropius, *Scope of Total Architecture*, p125.

34. H Ford, *My Life and Work* (1922, 1931), p19. See FW Taylor, *The Principles of Scientific Management* (New York, 1911). See also chapter two and B Russell, *Building Systems, Industrialisation and Architecture* (1981), pp85–95. For a recent study, see R Batchelor, *Henry Ford: Mass Production, Modernism and design* (Manchester, 1994).

35. CIAM, 'La Sarraz Declaration' (1928), in U Conrads, *Programmes and Manifestoes on 20th-century Architecture* (1970), p110.

36. Le Corbusier, *Towards a New Architecture* (1927), p246.

37. The article, published in Martin Wagner's journal *Wohnungswirtschaft* vol 1 nos 17/18 (1924) p157, is quoted in K Wilhelm, 'From the Fantastic to Fantasy', *Architectural Association Quarterly* 11 no 1 (1979) p7.

38. Gropius, in *Die Form* 2 (1927), pp275–7, translated as 'How can we build cheaper, better, more attractive houses?' in T and C Benton with D Sharp (eds), *Form and Function: A Source Book for the History of Architecture and Design 1890–1939* (1975), pp195–6.

39. E May, 'Mechanisierung des Wohnungbaues', *Das neue Frankfurt* 1 no 2 (1926) pp33–9; Wilhelm, *op cit*, pp7–8; Bullock, *op cit*, p340.

40. W Gropius, *The New Architecture and the Bauhaus* (1935), pp38–9 and 43.

41. Gropius, in Bentons and Sharp, *op cit*, p196. See Henry Ford on pre-production research: 'I do not believe in starting to make until I have discovered the best possible thing…. (Do) not even try to produce an article until you have satisfied yourself that utility, design and material are the best. If your researches do not give you that confidence, then keep on searching….' (Ford, *op cit*, p16).

42. E May, 'Unwin as planner for social welfare', *Town and Country Planning* 31 no 11 (November 1968) p428.

43. This point is worth emphasising since, following Tafuri (*Modern Architecture*, p246), the CIAM architects are often accused of technological fetishism. See May and Gropius on this, in Congrès Internationaux d'Architecture Moderne/Internationale Kongresse für Neues Bauen (CIAM), *Die Wohnung für das Existenzminimum* (Stuttgart, 1930), English summaries pp6 and 16.

44. Cf the title of *Cottage Plans and Common Sense* (Fabian Tract 109, 1902).

45. E May, 'Die Wohnung für das Existenzminimum', in CIAM, *op cit* (1930), English summaries p7.

46. *ibid*, p7.

47. Bauer, *op cit*, p200.

48. V Bourgeois, 'L'organisation de l'habitation minimum' in CIAM, *op cit* (1930), English summaries p10.

49. Gropius, in Bentons and Sharp, *op cit*, p195.

Chapter five – Sellier and Unwin

1. This paper was written for the Sellier centenary conference held at Suresnes in November 1983: see K Burlen (ed), *La Banlieue Oasis: Henri Sellier et les cités-jardins, 1900–1940* (Paris, 1987). For a good account of Sellier in English, see J Read, 'The Garden City and the Growth of Paris', *Architectural Review* 163 no 976 (June 1978) pp344–52. Also M Swenarton, 'Banlieue, Municipalités et Réformisme', *Planning History Bulletin* 6 no 1 (1984) pp4–6.

2. H Sellier, *La Crise du Logement et l'Intervention Publique en Matière d'Habitation Populaire dans l'Agglomération Parisienne* (Paris, 1921), pp253–93.

3. R Unwin, *Town Planning in Practice* (1909); PP 1918 Cd 9191 vii, 'Report of the Committee appointed by the president of the Local Government Board and the Secretary for Scotland to consider questions of building construction in connection with the provision of dwellings for the working classes in England and Wales, and Scotland, and report upon methods of securing economy and despatch in the provision of such dwellings' (the Tudor Walters Report). For the international prewar town planning movement, see A Sutcliffe, *Towards the Planned City: Germany, Britain, the United States and France, 1780–1914* (Oxford, 1981), chapter six. For developments in British housing during World War One, see M Swenarton, *Homes fit for Heroes. The Politics and Architecture of Early State Housing in Britain* (1981), chapter five. For European housing in the inter-war period see International Labour Office, *European Housing Problems since the War* (Geneva, 1924) and *Housing Policy in Europe* (Geneva, 1930); also C Bauer, *Modern Housing* (Boston, 1934).

4. Sellier, *op cit*, pp265–6.

5. Sellier, *op cit*, p258.

6. Sellier, *op cit*, p260.

7. Sellier, *op cit*, p273.

8. Sellier, *op cit*, pp282–3.

9. Sellier, *op cit*, pp285–6. Cf chapter ten of *Town Planning in Practice*: 'Of Buildings, and How the Variety of Each must be dominated by the Harmony of the Whole'.

10. Sellier, *op cit*, p256.

11. Sellier, *op cit*, pp257–8.

12. Sellier, *op cit*, pp262–3.

13. Sellier, *op cit*, p291.

14. Sellier, *op cit*, p287.

15. Sellier, *op cit*, p288.

Chapter six – CIAM, Teige and the *Existenzminimum*

1. Teige's paper was published in Congrès Internationaux d'Architecture Moderne/ Internationale Kongresse für Neues Bauen (CIAM), *Rationelle Bebauungsweisen*. (Frankfurt, 1931), pp64–70. An English abstract appeared in 'Abstract of Papers at the 3rd International Congress at Brussels of the International Committee for the Solution of the Problems in Modern Architecture', *Housing Study Guild Translations* 1 no 1 (New York, 1935), pp12–13. The full paper was published for the first time in English in 1987 in a translation by Christiane Crasemann Collins and Mark Swenarton, to which this essay provided the introduction: K Teige, 'The Housing Problem of the Subsistence Level Population: Summary of the National Reports at the International Congress for New Building (CIAM), 1930', *Habitat International* 11 no 3 (1987), pp147–151. Thanks are due to the RIBA for financial assistance with this translation. Since this essay was written, knowledge of Teige in the English-speaking world has been transformed by the work of Eric Dluhosch. See E Dluhosch and R Svácha (eds), *Karel Teige 1900–1951: L'Enfant Terrible of the Czech Modernist Avant-Garde* (Cambridge, Mass., 1999); and K Teige (trans. and with an introduction by E Dluhosch), *The Minimum Dwelling* (Cambridge, Mass., 2002). Recent additions to the literature on European social housing include N Stieber, *Housing Design and Society in Amsterdam: Reconfiguring Urban Order and Identity, 1900–1920* (Chicago, 1998) and E Blau, *The Architecture of Red Vienna 1919–1934* (Cambridge, Mass., 1999). See also E Mumford, *The CIAM Discourse on Urbanism, 1928–1960* (Cambridge, Mass., 2000).

2. International Labour Office (ILO), *European Housing Problems since the War* (Geneva, 1924), pp7 and 15.

3. *ibid*, p13.

4. *ibid*, pp17–40; N Bullock, 'Housing in Frankfurt 1925 to 1931 and the new *Wohnkultur*', *Architectural Review* 163 no 976 (June 1978) pp335–6; M Swenarton, *Homes fit for Heroes: The Politics and Architecture of Early State Housing in Britain* (1981), pp69–71.

5. A Ellinger, 'Socialisation Schemes in the German Building Industry', *International Labour Review* 1 (March 1921) pp287–301.

6. Swenarton, *op cit*, pp77–87.

7. H Searing, 'Amsterdam South: Social Democracy's Elusive Housing Ideal', *VIA* 4 (1980) p59.

8. CA Gulick, *Austria from Hapsburg to Hitler* (2 vols, Berkeley, 1948), vol 1 pp407–504.

9. ILO, *Housing Policy in Europe* (Geneva, 1930), pp44–5.

10. Gulick, *op cit*, chapters 9 and 14.

11. Bullock, *op cit*, p336.

12. International Housing Association, *Housing in Switzerland and in Frankfurt-a-M.* (Stuttgart, nd, 1932–3), p70.

13. ILO, *op cit* (1930), pp24 and 197–201; CM Bauer, *Modern Housing* (Boston, 1934), pp290–2; T Benton, 'Le Corbusier and the Loi Loucheur', *AA Files* 7 (1984) pp54–60.

14. Parliamentary Papers 1918 Cd 9191 vii, 'Report of the Committee appointed by the President of the Local Government Board and the Secretary for Scotland to consider questions of building construction in connection with the provision of dwellings for the working classes in England and Wales, and Scotland, and report upon methods of securing economy and despatch in the provision of such dwellings' (the Tudor Walters Report), para 349; Swenarton, *op cit*, pp140–1.

15. Reichsforschungsgesellschaft für Wirtschaftlichkeit im Bau- und Wohnungswesen e.V., *Technische Tagung in Berlin, 15 bis 17 April 1929* (Berlin, 1929); ILO, *op cit* (1930), pp367–8.

16. C Crasemann Collins, 'Concerned Planning and Design: the Urban Experiment of Germany in the 1920s', in FD Hirschbach and others (eds), *Germany in the Twenties: the Artist as Social Critic* (Minneapolis, 1980), pp30–47.

17. M Franciscono, *Walter Gropius and the Creation of the Bauhaus in Weimar: The Ideals and Artistic Theories of its Founding Years* (Urbana, 1971), pp72–3; B Miller Lane, *Architecture and Politics in Germany 1918–1945* (Cambridge, Mass., 1968), pp41–68; Bullock, *op cit*, pp335–42.

18. U Conrads, *Programmes and Manifestos on 20th-century Architecture* (1970), p110.

19. N Bullock, 'First the Kitchen – then the Façade', *AA Files* 6 (1984) pp59–67.

20. F Block, *Probleme des Bauens: der Wohnbau* (Potsdam, 1928).

21. K Wilhelm, 'From the Fantastic to Fantasy', *Architectural Association Quarterly* 11 (1979) pp6–8; M McLeod, 'Architecture or Revolution: Taylorism, Technocracy and Social Change', *Art Journal* 43 (1983) pp132–47; R Wiedenhoeft, *Berlin's Housing Revolution. German Reform in the 1920s* (Ann Arbor, 1985), pp33–51.

22. M Steinmann, 'Political Standpoints in CIAM 1928–1933', *Architectural Association Quarterly* 4 (1972) pp49–55; T Hilpert, 'Una polemica sul funzionalismo: Tiege (sic) – Le Corbusier 1929', *Casabella* 44 no 463–4 (November-December 1980) pp20–7; G Ciucci, 'The Invention of the Modern Movement', *Oppositions* 24 (1981) pp68–91.

23. Congrès Internationaux d'Architecture Moderne/Internationale Kongresse für Neues Bauen (CIAM), *Die Wohnung für das Existenzminimum* (Stuttgart, 1930), pp43–4. On Le Corbusier's South American trip, see C Crasemann Collins, 'Le Corbusier's Maison Errázurriz: A Conflict of Fictive Cultures', *Harvard Architectural Review* 6 (1987) pp38–53.

24. CIAM, *op cit* (1930), pp5–7.

25. K Teige (trans. O Mácel), 'Realismus und Formalismus' (excerpt from 'Der Formalismus', 1950–1951), *Archithese* 19 (1979) pp49–50; K Teige (with conclusion by P Kruntorad), *Liquidierung der 'Kunst'; Analysen, Manifeste* (Frankfurt, 1968); H Deluy, *Prague poésie; Front gauche; Karel Teige, le nouvel art proletarien, Nezval, Halas, Seifert, Jakobson, le mot 'structural', Maïakovski* (Paris, 1972), pp21–3. See also endnote 1 above.

26. Teige, *op cit* (1987), p150.

27. C Borngräber, 'Foreign Architects in the USSR', *Architectural Association Quarterly* 11 (1979) p40; Bullock, *op cit* (1978), p341.f

28. Gulick, *op cit*, p455; ILO, *op cit* (1930), p36.

29. Teige, *op cit* (1987), p148. See also F Engels, 'The Housing Question' (1872), in K Marx and F Engels, *Selected Works* (3 vols, Moscow, 1969–1970), vol 2 pp295–375.

30. Teige, *op cit* (1987), p149. See also IHA, *op cit* (1932–1933), p71; M Stratmann, 'Housing Policies in the Weimar Republic', *Architectural Association Quarterly* 11 (1979) p22; DP Silverman, 'A Pledge Unredeemed: the Housing Crisis in Weimar Germany', *Central European History* 3 (1970) p133; M Bowley, *Housing and the State, 1919–1944* (1945), pp45–7; Gulick, *op cit*, pp491–9; Benton, *op cit*, p60.

31. Borngräber, *op cit*, pp52 and 56; A Saint, *The Image of the Architect* (New Haven, 1983) p128.

32. Le Corbusier (trans. A Eardley, with introduction by J Giraudoux), *The Athens Charter* (New York, 1973), pp1–38; *Parametro* 52, special issue, 'From Brussels to Athens: the Functional City' (December 1976) p48.

Chapter seven – The education of an urbanist

1. 'The Architect's Contribution. The Inaugural Address by the President, Dr Raymond Unwin, read before the RIBA on Monday 2 November 1931', *RIBA Journal* 3rd ser., 39 no 1 (November 1931) p9. I am particularly indebted to Mervyn Miller for making available to me his copied set of Unwin's correspondence of the 1880s from the Hitchcock Collection. On Unwin, see MK Miller, *To Speak of Planning is to Speak of Unwin. The Contribution of Sir Raymond Unwin (1863–1940) to the Evolution of British Town Planning* (University of Birmingham PhD, 1981), which largely supersedes the earlier published studies: W Creese, *The Search for Environment. The Garden City: Before and After* (1966); MG Day, 'The Contribution of Sir Raymond Unwin (1863–1940) and R Barry Parker (1867–1947) to the Development of Site Planning Theory and Practice, c1890–1918', in A Sutcliffe (ed), *British Town Planning: the formative years* (Leicester, 1981) pp156–99; and F Jackson, *Sir Raymond Unwin. Architect, Planner, Visionary* (1985). Since publication of this essay, Mervyn Miller's thesis has been published: M Miller, *Raymond Unwin: Garden Cities and Town Planning* (Leicester, 1996). Other contributions to the literature on the garden city movement include F Bollerey, G Fehl, K Hartmann, *Im Grünen wohnen – im Blauen planen: Ein Lesebuch zur Gartenstadt* (Hamburg, 1990); P Hall and C Ward, *Sociable Cities: the legacy of Ebenezer Howard* (Chichester, 1998); M Harrison, *Bournville: Model Village to Garden Suburb* (Chichester, 1999); and M Miller, *Hampstead Garden Suburb: Arts and Crafts Utopia?* (Chichester, 2006). For a recent view of the utopian strand from the garden city movement to the present, see D Pinder, *Visions of the City: Utopianism, Power and Politics in Twentieth-Century Urbanism* (Edinburgh, 2006).

2. 'The Royal Gold Medal. Presentation to Sir Raymond Unwin', *RIBA Journal* 3rd ser., 44 no 12 (April 1937) p582. See also UMDAP Unwin Papers, 'Notes for speech to London Society' (nd); 'Founder's Day Ceremony: Sir Raymond Unwin and Planning', *Manchester Guardian* (16 May 1935) p13.

3. *Architectural Review* 163 no 976 (June 1978), 'The Garden City Idea' (special issue); above, chapter four.

4. 'Edward Carpenter and "Towards Democracy"', in G Beith (ed), *Edward Carpenter. In Appreciation* (1931) p234.

5. *Daily Independent* (Sheffield, 5 December 1932) p6; Miller, *op cit* (1981), p53.

6. 'The Prince and his Hand', *Commonweal* 5 no 157 (12 January 1889) p10.

7. 'Co-operation and Competition', *Commonweal* 2 no 26 (10 July 1886) p114.

8. Beith, *op cit*, p234.

9.   PC, Unwin to Ethel Parker, 4 May [1884].

10.   Beith, *op cit*, p235.

11.   PC, Unwin to Ethel Parker, 10 January 1891. In the surviving 17 letters from the period 1884–1891, Carpenter figures in ten.

12.   Beith, *op cit*, pp234–5.

13.   UMDAP, Unwin Papers, Notes for lecture on Carpenter, 3 July 1939.

14.   Beith, *op cit*, pp234–6.

15.   UMDAP, Unwin Papers, Notes for lecture on Carpenter, 3 July 1939.

16.   UMDAP, Unwin papers, 1887 Diary, entry for 12 August. The diary, comprising entries addressed to Ethel Parker, substituted for correspondence between them, which had been prohibited by her father.

17.   E Carpenter, *My Days and Dreams* (1916; 3rd edn 1921), pp139–40 and p114.

18.   PC, Unwin to Ethel Parker, nd [3 May 1885]; UMDAP, Unwin Papers, 1887 Diary, entry for 11 September.

19.   PC, Unwin to Ethel Parker, nd [3 May 1885].

20.   BAL, Unwin Papers, UN 15/2, 'The Dawn of a Happier Day' (January 1886) p26.

21.   'Socialist Tactics. A Third Course', *To-Day* ns 7 no 49 (December 1887) p185; UMDAP, Unwin Papers, 1887 Diary, entry for 19 May.

22.   'The Question of Political Policy. The Movement Past and Present', *Labour Leader* (18 January 1902) p21.

23.   PC, Unwin to Ethel Parker, nd [31 January 1885].

24.   IISH, Socialist League Papers 3031, Unwin to the Secretary, 21 August 1885.

25.   IISH, Socialist League Papers 606, Monthly Reports October 1885–March 1886.

26.   IISH, Socialist League Papers 3042, Unwin to the Secretary, 27 June 1886.

27.   *Commonweal* 2 no 44 (13 November 1886) p264; EP Thompson, *William Morris: Romantic to Revolutionary* (1955; revised edn, 1977) p414.

28.   UMDAP, Unwin Papers, 1887 Diary, entry for 16 May.

29.   IISH, Socialist League Papers 3042, Unwin to the Secretary, 27 June 1886.

30.   IISH, Socialist League Papers 3042, Unwin to the Secretary, 4 July 1886.

31.   'The Axe is Laid unto the Root', *Commonweal* 2 no 31 (14 August 1886) p155.

32.   PC, Unwin to Ethel Parker, 10 January and 20 January 1891.

33.   UMDAP, Unwin Papers, 1887 Diary, entry for 18 September.

34.   C Tsuzuki, *Edward Carpenter 1844–1929. Prophet of Human Fellowship* (Cambridge, 1980) p66.

35.   *ibid*.

36.   Carpenter, *op cit*, pp131–2.

37.   Tsuzuki, *op cit*, p96.

38. PC, Unwin to Ethel Parker, 26 January 1891.

39. JL Mahon, quoted in Thompson, *op cit*, p473.

40. UMDAP, Unwin Papers, 1887 Diary, entry for 6 June.

41. *Commonweal* 3 no 80 (23 July 1887) p240.

42. UMDAP, Unwin Papers, 1887 Diary, entry for 2 August.

43. IISH, Socialist League Papers 3046, Unwin to the Secretary, 18 September 1887.

44. 'Socialist Tactics. A Third Course', *To-Day* ns 7 no 49 (December 1887) pp180–186.

45. 'Westward Ho!', *Commonweal* 6 no 227 (10 May 1890) p151.

46. UMDAP, Unwin Papers, 1887 Diary, entry for 16 August.

47. UMDAP, Unwin Papers, 1887 Diary, entry for 2 August.

48. PC, Unwin to Ethel Parker, 4 May [1884].

49. UMDAP, Unwin Papers, 1887 Diary, entry for 3 August.

50. UMDAP, Unwin Papers, 1887 Diary, entry for 21 August.

51. UMDAP, Unwin Papers, 1887 Diary, entry for 22 May.

52. UMDAP, Unwin Papers, 1887 Diary, entry for 23 August.

53. UMDAP, Unwin Papers, 1887 Diary, entry for 5 May to 15 June, passim; 'Positivism and Socialism', *Commonweal* 3 no 80 (23 July 1887) p235.

54. PC, Unwin to Ethel Parker, 10 January [1891].

55. PC, Unwin to Ethel Parker, 24 April 1891.

56. PC, Unwin to Ethel Parker, 9 August 1891.

57. *Labour Leader* (18 January 1902) p21.

58. *Labour Annual* (1897) p245.

59. *Labour Prophet* 6 no 74 (February 1898) p159; B Parker, 'Memoir of Sir Raymond Unwin', *RIBA Journal* 3rd ser., 47 no 9 (15 July 1940) p209.

60. 'The Place of the Labour Church', *Labour Prophet and Labour Church Record* 6 no 75 (March 1898) pp161–2.

61. Parker, *op cit*, p209.

62. UMDAP, Unwin Papers, Notes for lecture on Hinton, nd [1905–6].

63. 'Early Communal Life and What It Teaches. Pt V', *Commonweal* 3 no 70 (14 May 1887) p157.

64. BAL, Unwin Papers UN 15/4, 'Gladdening v. Shortening the Hours of Labour', pp15–16.

65. 'The Passing of Sir Raymond Unwin', *Labour's Northern Voice* (August 1940) p2.

66. (with B Parker) *The Art of Building a Home* (1901), p61.

67. 'The Art of Building a Home', *Architects' Magazine* 1 no 12 (October 1901) p224.

68. *The Art of Building a Home*, p1.

69.   *ibid*, p111.

70.   *Cottage Plans and Common Sense* (Fabian Tract 109, 1902) p3.

71.   *The Art of Building a Home*, p114. For Ruskin's exhortation to 'go to Nature' see J Ruskin, *Modern Painters* I (1843) in ET Cook and A Wedderburn (eds), *The Works of John Ruskin* (39 vols, 1903–1912) vol 3 p624.

72.   *ibid*, p64.

73.   *ibid*, p66.

74.   *Architects' Magazine* 1 no 4 (February 1901) p68.

75.   *The Art of Building a Home*, p132.

76 . UMDAP, Unwin Papers, 1887 Diary, entry for 21 August.

77.   'Sutton Hall', *Commonweal* 5 no 179 (15 June 1889) p190.

78.   *Labour Prophet and Labour Church Record* 6 no 63 (March 1897) p46. See also WHG Armytage, *Heavens Below. Utopian Experiments in England 1560–1960* (1961) chapter six; and S Pierson, *Marxism and the Origins of British Socialism. The Struggle for a New Consciousness* (Ithaca, 1973) p221.

79.   C Lee, 'From a Letchworth Diary', *Town and Country Planning* 21 no 113 (September 1953) pp435–6. For Unwin's appointment as architect at Letchworth Garden City, see Miller, *op cit* (1981), pp232–42.

80.   'Co-operation in Building', *Architects' Magazine* 1 no 2 (December 1900) p20 and 1 no 3 (January 1901) pp37–8.

81.   BAL, Unwin Papers UN 15/2, 'The Dawn of a Happier Day' (1886) pp1–3.

82.   *The Art of Building a Home*, p92.

83.   *ibid*, p100.

84.   *ibid*, p104.

85.   'The Houses our Forefathers Lived in', *Architects' Magazine* 1 no 3 (January 1901) p44.

86.   *The Art of Building a Home*, pp104–5.

87.   'Of the Building of Houses in Garden City', *The Garden City Conference at Bourneville. Report of Proceedings* (1901) p72.

88.   FGCHM, 'The Improvement of Towns. A paper read at the Conference of the National Union of Women Workers of Great Britain and Ireland, November 8th 1904, by Mr Raymond Unwin' (offprint, 1904) p8.

89.   *Labour Chronicle* 1 (Leeds, May 1893) p7. On the ILP see Pierson, *op cit*, pp268–71.

90.   'The Question of Political Policy. The Movement Past and Present', *ILP News* 5 no 58 (January 1902) pp1–2.

91.   *ibid*, p4.

92.   'The Question of Political Policy. The Movement Past and Present', *Labour Leader* (18 January 1902) p21; *Archives of the Independent Labour Party. Series III. The Frances Johnson Correspondence, 1888–1950. Part 1 1888–1908* (microfiche, Hassocks, 1980), reel 5 (1902), 12A, Unwin to K Hardie, 26 January 1902.

93. *Cottage Plans and Common Sense*, p4.

94. *ibid*, p2. On Unwin and post-war housing see M Swenarton, *Homes fit for Heroes. The Politics and Architecture of Early State Housing in Britain* (1981) pp94–5.

95. *Cottage Plans and Common Sense*, p2.

96. *ibid*, p3.

97. *ibid*, p13.

98. *ibid*, p15.

99. NCO, Fabian Society Papers, Membership Record Cards, 'Raymond Unwin'.

100. *Labour Annual* (1903) p79.

101. FGCHM, (with B Parker) 'Cottages near a Town' (offprint, Northern Art Workers Guild, 1903) p12.

102. 'City Planning', *Cambridge Independent Press* (16 February 1906) p3.

103. Miller, *op cit* (1981), p184.

104. BAL, Unwin Papers UN 15/5/4, Quotations from Carpenter transcribed by Unwin.

105. UMDAP, Unwin Papers, Notes for lecture on Carpenter (3 July 1939).

106. 'City Planning', p3.

107. BAL, Unwin Papers UN 2/1, Notes for lecture on Town Planning (28 February 1908).

108. 'The Improvement of Towns', pp2 and 15.

109. *ibid*, p8.

110. 'Early Communal Life and What It Teaches. Pt II', *Commonweal* 3 no 67 (23 April 1887) p135.

111. 'The Improvement of Towns', pp7–8.

112. *ibid*, p8.

113. UMDAP, Unwin Papers, Notes for speech on Aristocracy and Democracy, nd; PC, Unwin to Ethel Parker, 5 December 1885.

114. PC, Unwin to Ethel Parker, nd [November 1885].

115. 'Looking Back and Forth. An Address to Students by the President, Dr Raymond Unwin', *RIBA Journal*, 39 no 6 (January 1932) pp205–6. For Ruskin's ideas about Gothic labour and Morris' re-working of them see M Swenarton, *Artisans and Architects: the Ruskinian tradition in architectural thought* (1989), pp1–31 and 61–95.

116. BAL, Unwin Papers UN 15/4, 'Gladdening v. Shortening the Hours of Labour' (1897).

117. *The Art of Building a Home*, p88.

118. 'WR Lethaby: An Impression and a Tribute', *RIBA Journal* 3rd ser., 39 no 8 (20 February 1932) p304.

119. *Labour Leader* (18 January 1902) p21.

120. 'Some Objections Answered', *Commonweal* 4 no 126 (9 June 1888) p181.

121. 'The Saving of Labour', *Commonweal* 4 no 124 (26 May 1888) p163.

122. *Commonweal* 4 no 126 (9 June 1888) p181.

Chapter eight – Unwin and Sitte

1.   GR Collins and C Crasemann Collins, *Camillo Sitte: The Birth of Modern City Planning* (New York, 1986) p363; M Miller, 'Der rationelle Enthusiast: Raymond Unwin als ein Bewunderer deutschen Städtebaus', *Bauwelt* 36 (24 September 1982) p320. For an international perspective on Sitte, see G Zucconi, *Camillo Sitte e i suoi Interpreti* (Milan, 1992). Although Unwin's library included a substantial German collection (the bibliography to *Town Planning in Practice* cited the 1909 German edition of Sitte as well as the French version) his German apparently did not allow him to read Sitte's original text. See M Miller, *Hampstead Garden Suburb: Arts and Crafts Utopia?* (Chichester, 2006), p52.

2.   Collins, *op cit*, p78.

3.   A Sutcliffe, *Towards the Planned City: Germany, Britain, the United States and France 1870–1914* (Oxford, 1981), pp47–87 and pp163–201.

4.   TC Horsfall, *The Improvement of the Dwellings and Surroundings of the People: The Example of Germany* (Manchester, 1904), p19. See also GR Searle, *The Quest for National Efficiency: A Study in British Politics and Political Thought 1899–1914* (1971).

5.   W Ashworth, *The Genesis of Modern British Town Planning* (1954), p169.

6.   M Swenarton, *Homes fit for Heroes: The Politics and Architecture of Early State Housing in Britain* (1981) pp32–4 and 38–44. On the garden city movement see W Creese, *The Search for Environment. The Garden City: Before and After* (New Haven, 1966) and endnote 1 to chapter seven above.

7.   R Unwin, 'The Question of Political Policy. The Movement Past and Present', *Labour Leader* (18 January 1902) p21.

8.   R Unwin, *Town Planning in Practice* (1909), p13.

9.   R Unwin, *Cottage Plans and Common Sense* (Fabian Tract 109, 1902), p2.

10. Unwin, *op cit* (1909), p138.

11. R Unwin, 'The Planning of Residential Districts of Towns', *Transactions of the VII International Congress of Architects, London 1906* (1908), p420.

12. R Unwin, 'City Planning. Lecture at Cambridge by the Architect of Garden City', *Cambridge Independent Press* (16 February 1906) p3.

13. Unwin, *op cit* (1909), p225.

14. Collins, *op cit*, p230.

15. Unwin, *op cit* (1909), p318.

16. Unwin, *op cit* (1909), p98.

17. Unwin, *op cit* (1909), p104.

18. Unwin, *op cit* (1909), p270. See also ibid, p52, and Collins, *op cit*, p98.

19. Unwin, *op cit* (1909), p194.

20. Unwin, *op cit* (1909), p197.

21. Unwin, *op cit* (1909), p197.

22. Unwin, *op cit* (1909), p197.

23. Unwin, *op cit* (1909), p198.

24. Unwin, *op cit* (1909), p207.

25. Unwin, *op cit* (1909), pp221–2.

26. Unwin, *op cit* (1909), p223.

27. Unwin, *op cit* (1909), p245.

28. Unwin, *op cit* (1909), pp245–6.

29. Unwin, *op cit* (1909), p246.

30. Collins, *op cit*, p230.

31. Unwin, *op cit* (1909), p249.

32. Unwin, *op cit* (1909), p220. Also Collins, *op cit*, p94.

Chapter nine – Rammed earth revival

1. C Williams-Ellis, *Cottage Building in Cob, Pisé, Chalk and Clay: A Renaissance* (1919); Department of Scientific and Industrial Research, *Experimental Cottages: A Report on the Work of the Department at Amesbury, Wilts, by WR Jaggard, FRIBA* (1921); Building Research Board, *Special Report no 5: Building in Cob and Pisé de Terre* (1922).

2. English Heritage, *Terra 2000. 8th International Conference on the study and conservation of earthen architecture, Torquay, Devon, UK, May 2000: Preprints* (2000).

3. O Kapfinger, *Martin Rauch: Rammed Earth* (Basel, 2001) pp82–94; 'Eden regained: Nicholas Grimshaw & Partners in Cornwall', *Architecture Today* 119 (June 2001) pp44–58.

4. P Walker, R Keable, J Martin and V Maniatidis, *Rammed earth: design and construction guidelines* (Watford, 2005).

5. Private communication. See also J Keable, *Rammed Earth Structures: a code of practice* (1996).

6. English Heritage, *op cit*, p xii.

7. C Williams-Ellis with an introduction by GT Pearson, *Building in cob, pisé and stabilized earth* (Shaftesbury, 1999).

8. C Williams-Ellis, J [Eastwick-Field] and E Eastwick-Field, *Building in cob, pisé and stabilized earth* (1947), pp103–12.

9. R Eaton, 'Mud: an examination of earth architecture', *Architectural Review* 170 no 1016 (October 1981) p227; GT Pearson, *Conservation of Clay and Chalk Buildings* (Shaftesbury, 1992), pp18 and 22–24.

10. RJ Curry and S Kirk, *Philip Webb in the North* (Middlesbrough, 1984), p27; D O'Neill, *Lutyens Country Houses* (1980), pp45–9 and 51–8; M Richardson, *Architects of the Arts and Crafts Movement* (1983), pp54–5; G Hoare and G Pyne, *Prior's Barn & Gimson's Coxen: Two Arts and Crafts Houses* (Budleigh Salterton, 1978), np. Since this essay was written, Sheila Kirk's magisterial study of Webb has eclipsed earlier studies: see S Kirk, *Philip Webb: Pioneer of Arts and Crafts Architecture* (Chichester, 2005), p128.

11. BRB, *op cit* (1922), p29.

12. Williams-Ellis, *op cit* (1919), p5.

13. A Williams-Ellis, *All Stracheys are Cousins: Memoirs* (1983), p18; A Strachey, *St Loe Strachey: His Life and His Paper* (1930), p183.

14. A Williams-Ellis, *op cit*, p19.

15. *ibid*, p20.

16. J St L Strachey, *The Adventure of Living: A Subjective Autobiography* (1922), p160.

17. *ibid*, p402.

18. Williams-Ellis, *op cit* (1919), p13.

19. National Unionist Association of Conservative and Liberal Unionist Organisations, *The History of Housing Reform* (1913), pp20–2; L Weaver, *The Country Life Book of Cottages, costing from £150 to £600* (1913) pp20–21.

20. A Strachey, *op cit*, p186.

21. *ibid*, p187; J Cornes, *Modern Housing in Town and Country* (1905), pp123–96.

22. Weaver, *op cit*, p21; W Thompson, *Housing Up-to-Date* (1907), p155.

23. *The Garden City* ns 2 no 19 (August 1907) p466.

24. M Fraser, *John Bull's Other Homes: State Housing and British Policy in Ireland, 1883–1922* (Liverpool, 1996), p41; M Swenarton, *Homes fit for Heroes: The Politics and Architecture of Early State Housing in Britain* (1981), p32.

25. Swenarton, *op cit*, p33; Fraser, *op cit*, p120.

26. PP 1913 Cd 6708 xv, 'Report of the Departmental Committee appointed by the President of the Board of Agriculture and Fisheries to inquire into and report as to Buildings for Small Holding in England and Wales, together with Abstract of the Evidence, Appendices, and a series of Plans and Specifications'. The 1913 report was followed by the appointment in November 1913 of a further committee, again including Unwin, to make detailed recommendations on plans, specifications and building methods: see Advisory Committee on Rural Cottages, *Report of the Committee appointed by the President of the Board of Agriculture and Fisheries to consider and advise the Board on Plans, Models, Specifications and Methods of Construction for Rural Cottages and Outbuildings* (1915). Also Swenarton, *op cit*, pp41–2 and 44.

27. 'A Model Black Weather-Boarded Cottage', *The Spectator* 4438 (19 July 1913) pp91–2; 'The Model Cottage at Merrow', *The Spectator* 4440 (13 August 1913) pp171–2; A Williams-Ellis, *op cit*, pp41–2.

28. L Weaver, 'The Search for Cheapness', *Country Life* 34 no 878 (25 October 1913) pp556–7.

29. 'The Country Life National Competition for Cottage Designs', *Country Life* 34 no 883 (6 December 1913) pp834–835; Weaver, *op cit* (1913), pp6–16; Fraser, *op cit*, p117.

30. *The Spectator* 4456 (22 November 1913) p865.

31. Williams-Ellis, *op cit* (1919), p14; see also *The Spectator* 4457 (29 November 1913) p912, 4459 (13 December 1913) p1024, 4464 (17 January 1914) p92 and 4471 (7 March 1914) p388.

32. Williams-Ellis, *op cit* (1919), p16.

33. A Strachey, *op cit*, p294.

34. Williams-Ellis, *op cit* (1919), p19.

35. PP 1918 Cd 9144 ix, 'Report of the Committee of the Privy Council for Scientific and Industrial Research for the Year 1917–18', p47.

36. C Williams-Ellis, *Around the World in Ninety Years* (Portmeirion, 1978), p126; see also R Haslam, *Clough Williams-Ellis* (1996), pp6–22.

37. C Williams-Ellis, *Lawrence Weaver* (1933), pp36–7 and 40.

38. A Williams-Ellis, *op cit*, pp42–44; J Jones, *Clough Williams-Ellis, The Architect of Portmeirion: A Memoir* (Bridgend, 1996), pp42–3; A Strachey, *op cit*, p289; C Williams-Ellis, 'A Hundred-Guinea Cottage', *The Spectator* 4469 (21 February 1914) pp296–7; C Williams-Ellis, 'The Model Hundred-Guinea Cottage at Merrow' *The Spectator* 4482 (23 May 1914) pp863–4.

39. FLC Floud, 'Land Settlement', *Journal of the Board of Agriculture* 26 no 7 (October 1919) p676; Swenarton, *op cit*, pp67–87.

40. 'Estimate of Expenditure under the Land Settlement (Facilities) Bill', *Journal of the Board of Agriculture* 26 no 2 (May 1919) p192. See also EJT Collins (ed), *The Agrarian History of England and Wales vol 7 1850–1914 Part I* (Cambridge, 2000), pp780–5.

41. Ministry of Agriculture and Fisheries, *Report of Proceedings under the Small Holdings Colonies Acts 1916 and 1918 for the Period ended 31st March 1920* (1921), p.2. See also EH Whetham, *The Agrarian History of England and Wales, volume 8 1914–1939* (Cambridge, 1978), pp85–8 and A Howkins, *The Death of Rural England: A social history of the countryside since 1900* (2003), p88.

42. See Lloyd George's speech to the War Cabinet, NA, CAB 23/9, WC 539 (3 March 1919), as discussed in PB Johnson, *Land Fit for Heroes. The Planning of British Reconstruction 1916–1919* (Chicago, 1968), pp347–51; Swenarton, *op cit*, p77–81; Fraser, *op cit*, p298.

43. *Journal of the Board of Agriculture* 25 no 11 (February 1919) p1369; Ministry of Agriculture and Fisheries, *Land Settlement in England & Wales: Being a Report of Proceedings under the Small Holdings & Allotments Acts 1908 to 1919 for the Period 1919 to 1924* (1925), pp36–45; Swenarton, *op cit*, pp80–1.

44. 'Land Settlement in the Mother Country', *Journal of the Board of Agriculture* 25 no 10 (January 1919) p1157; Johnson, *op cit*, p348.

45. Williams-Ellis, *op cit* (1933), p46; LT Weaver, *Lawrence Weaver 1876–1930: an annotated bibliography* (York, 1989), pp13–14; Johnson, *op cit*, p348.

46. NA, MAF 39/36, Board of Agriculture and Fisheries, List of Officers, their Rank and Salaries (1 April 1919), p1.

47. Ministry of Agriculture, *Land Settlement in England & Wales* (1925), p15. See also PP 1918 Cd 9191 vii, 'Report of the Committee appointed by the President of the Local Government Board and the Secretary for Scotland to consider questions of building construction in connection with the provision of dwellings for the working classes in England and Wales, and Scotland, and to report upon methods of securing economy and despatch in the provision of such dwellings' (the Tudor Walters Report).

48. Board of Agriculture and Fisheries, *Manual for the Guidance of County Councils and*

their Architects in the Equipment of Small Holdings, Part I, The Planning and Construction of Cottages (1919), pp9–10; Ministry of Agriculture, *Land Settlement in England & Wales* (1925), p15–16.

49. Ministry of Agriculture, *Report for the Period ended 31st March 1920*, p7.

50. 'Pisé and Other Cottages at Amesbury', *Country Life* 48 no 1246 (20 November 1920) p680.

51. Ministry of Agriculture and Fisheries, *Report of Proceedings under the Small Holdings Colonies Acts 1916 and 1918 for the Two Years 1921–22 and 1922–23* (1924), p18.

52. Cd 9144, pp47–8; DSIR, *Experimental Cottages at Amesbury*, Prefatory Note (np).

53. NA, DSIR 4/1, letter to Local Government Board (3 February 1919) and memorandum by Lord Curzon on proposed Building Research Board (23 April 1919).

54. DSIR, *Experimental Cottages at Amesbury*, Prefatory Note (np). See also AAH Scott, *Reinforced Concrete in Practice* (1915) and FE Drury and WR Jaggard, *Architectural Building Construction* (3 vols, Cambridge, 1916–1923).

55. 'Land Settlement of Ex-Service Men', *Journal of the Board of Agriculture* 25 no 10 (January 1919) pp1244 and 1245.

56. Ministry of Agriculture and Fisheries, *Land Settlement 1919 to 1924*, pp33–6; Swenarton, *op cit*, pp136–9.

57. Williams-Ellis, *op cit* (1933), p48.

58. NA, MAF 39/36, List of Officers (1 April 1919), p21.

59. Williams-Ellis, *op cit* (1919), p5.

60. *ibid*, pp28–9.

61. BAL, Jacqueline Tyrhwhitt Papers, 66/15, 'Amesbury' (2 September 1919), p1; Swenarton, *op cit*, p157. The cottage at Newlands Corner still exists, albeit much extended and clad in an outer brick skin.

62. 'Experimental Cottage Building', *Journal of the Ministry of Agriculture* 27 no 3 (June 1920) p217.

63. NA, MAF 48/306, Amesbury Farm Settlement, letter to the Treasury (17 April 1919).

64. BAL, Tyrwhitt Papers, 66/15, 'Amesbury Estate Building Works' (7 January 1920), p1. Thomas Tyrwhitt (1874–1956) had previously worked in private practice in Hong Kong and as a government architect in the Transvaal before returning to London in 1908. His best known work is the Indian Memorial Gateway to Brighton Pavilion (1920).

65. Ministry of Agriculture, *Report for the Period ended 31st March 1920*, p11.

66. C Williams-Ellis, 'The Modern Cottage: Experiments in Pisé at Amesbury', *Journal of the Ministry of Agriculture* 27 no 6 (September 1920) p530.

67. 'Experimental Cottage Building', Journal of the Ministry of Agriculture, p216.

68. Ministry of Agriculture and Fisheries, *Manual for the Guidance of County Councils and their Architects in the Equipment of Small Holdings, Part I, Planning and Construction of Cottages; Part II, Planning and Construction of Farm Buildings, 3rd Edition, Revised* (September 1920), p12; also Williams-Ellis, 'The Modern Cottage', *Journal of the Ministry of Agriculture*, p531.

69. Williams-Ellis, 'The Modern Cottage', *Journal of the Ministry of Agriculture*, p533.

70. L Weaver, 'Building for Land Settlement: A Survey of the Ministry's Work', *Journal of the Ministry of Agriculture* 28 no 2 (May 1921) p101.

71. DSIR, *Experimental Cottages at Amesbury*, p74.

72. BAL, Tyrhwhitt Papers, 66/15, 'Ministry of Agriculture and Fisheries, Amesbury Farm Settlement, Experimental Cottages' (30 September 1920), pp2–3.

73. NA, MAF 48/307, 'Amesbury Farm Settlement. Costs' (11 November 1922).

74. Cd 9144, p49. The Jaggard report presented figures on walling costs on a unit cube basis, in which earth materials came out better in relation to brick but not to concrete: DSIR, *Experimental Cottages at Amesbury*, p33.

75. DSIR, *Experimental Cottages at Amesbury*, p3; BRB, *Special Report no 5*, p21.

76. Williams-Ellis, 'The Modern Cottage', *Journal of the Ministry of Agriculture*, p534.

77. NA, DSIR 4/4, Building Research Board, Meetings 1920, minutes of third meeting (25 November 1920).

78. BRB, *Special Report no 5*, p2.

79. Weaver, 'Building for Land Settlement', p100; Ministry of Agriculture, *Land Settlement 1919 to 1924*, p31; Ministry of Agriculture, *Manual* (September 1920), p5.

80. Ministry of Agriculture, *Land Settlement 1919 to 1924*, p22.

81. NA, MAF 39/383, Ministry of Agriculture and Fisheries, List of Officers (1921); MAF 39/140, Ministry of Agriculture and Fisheries, Registers of Service, Established Staff 1913–1921, p198.

82. Ministry of Agriculture, *Land Settlement 1919 to 1924*, pp7, 14 and 53. For a more recent assessment see Whetham, *op cit*, pp137–9.

83. B Russell, *Building Systems, Industrialization, and Architecture* (1981), pp201–212; B Finnimore, *Houses from the Factory: System Building and the Welfare State 1942–1974* (1989).

84. SR Henderson, *The Work of Ernst May, 1919–1930* (Columbia University PhD, 1990), pp213–218; K Frampton, 'Architecture and critical regionalism', *RIBA Transactions 3* (1983) pp15–25.

85. US Department of Agriculture, *Farmers' Bulletin No 1500, Rammed Earth Walls for Buildings* (Washington DC, 1937).

86. Williams-Ellis, 'The Modern Cottage', *Journal of the Ministry of Agriculture*, p529.

87. Williams-Ellis and Eastwick-Field, *op cit* (1947). For a more recent discussion of cement see J Ramshaw, 'Matter of debate: is there a green future for concrete?', *EcoTech* 5 (May 2002) pp18–19.

Chapter ten – Breeze blocks and Bolshevism

1. Building Research Station, *Golden Jubilee Congress: Proceedings 9–15 June 1971* (1972), p211.

2. PB Johnson, *Land Fit for Heroes: The Planning of British Reconstruction 1916–1919* (Chicago, 1968); PR Wilding, *Government and Housing: A Study in the Development of Social Policy, 1906–1939* (University of Manchester DPhil, 1970); BB Gilbert, *British*

Social Policy 1914–1939 (1970); M Swenarton, *Homes fit for Heroes: The Politics and Architecture of Early State Housing in Britain* (1981); M Fraser, *John Bull's Other Homes: State Housing and British Policy in Ireland, 1883–1922* (Liverpool, 1996).

3. RB White, *Prefabrication: A history of its development in Great Britain* (1965); BREA, Building Research Station, Internal Note IN98/66, Information Division, *A short history of Building Research in the United Kingdom from 1917 to 1946 – in particular that of the Building Research Station, by RB White, with the collaboration of Station Officers* (1966); FM Lea, *Science and Building: A History of the Building Research Station* (1971); BRS, *op cit* (1972); HF Heath and AL Hetherington, *Industrial Research and Development in the UK* (1946); G Atkinson, 'Raymond Unwin: Founding Father of the BRS', *RIBA Journal* 3rd series 78 no 10 (October 1971), pp446–8.

4. PP 1917–1918 Cd 8696 xv, 'Summary of the Reports of the Commission of Enquiry into Industrial Unrest by the Rt Hon GN Barnes', pp6–7; Swenarton, *op cit*, pp70–2.

5. NA, CAB 23/3, WC 194, 24 July 1917; PP 1918 Cd 9191 vii, ' Report of the Committee appointed by the President of the Local Government Board and the Secretary for Scotland to consider questions of building construction in connection with the provision of dwellings for the working classes in England and Wales, and Scotland, and report upon methods of securing economy and despatch in the provision of such dwellings' (the Tudor Walters Report), p3. Also Swenarton , *op cit*, pp48–66; M Miller, *Raymond Unwin: Garden Cities and Town Planning* (Leicester, 1996), pp149–60.

6. NA, DSIR 3/51, memorandum on enquiry by R Unwin, 30 August 1917.

7. NA, DSIR 3/51, memorandum on Research in Building Materials and Methods of Construction by AS Barnes, September 1917.

8. *ibid.*

9. NA, DSIR 3/51, memorandum on enquiry by R Unwin, 30 August 1917.

10. NA, DSIR 3/51, Minute of Advisory Council, 3 October 1917. See also DSIR 3/51, note on interview with Sir Horace Munro, 29 June 1917.

11. NA, DSIR 3/51, Lyon to Unwin, 5 October 1917.

12. NA, DSIR 3/51, Heath to Tudor Walters, 1 October 1917.

13. MI Cole (ed), *Beatrice Webb's Diaries 1912–1924* (1952), p86; NA, DSIR 3/51, memorandum by AS Barnes, September 1917, and letters of invitation to Humphreys, Prior and Rowntree, October 1917.

14. Heath and Hetherington, *op cit*, pp282–283; M Bowley, *The British Building Industry: four studies in response and resistance to change* (Cambridge, 1966), p190; Lea, *op cit*, pp12–13.

15. See Swenarton, *op cit*, pp67–87.

16. NA, CAB 23/9, WC 539, Lloyd George, 3 March 1919; *Parliamentary Debates, Commons* 114 col 1956, 8 April 1919 (W Astor). See also Fraser, *op cit*, pp186–8; Swenarton, *op cit*, pp70–81 and pp117–19.

17. NA, DSIR 3/51, First Report of the BMRC, 3 January 1918.

18. NA, DSIR 3/51, Willis to Heath, 9 January 1918.

19. NA, DSIR 3/51, note by Lyons, 15 Jan 1918.

20. NA, DSIR 3/51, Earle to Heath, 23 January 1918.

21. NA, DSIR 3/51, Lloyd to Unwin, 28 January 1918.

22. NA, DSIR 3/51, minute by Heath, 10 April 1918.

23. NA, DSIR 3/51, Minutes of Advisory Council, 6 March 1918 and 5 June 1918.

24. NA, DSIR 3/50, BMRC, 11th meeting, 6 June 1918, and 12th meeting, 9 October 1918; PP 1920 Cmd 905 xxv, 'Report of the Committee of the Privy Council for Scientific and Industrial Research for the year 1919–1920', pp60–61; DSIR, *Report of the Building Research Board for the Period Ended 31st December 1926* (1927), p1. From 1921 research in timber was undertaken by the Forest Products Research Board, which was established in that year.

25. Department of Scientific and Industrial Research, Building Research Board, *Special Report no 2, Experiments on Floors* (1921); *Special Report no 3, The Stability of Thin Walls* (1921); *Special Report no 4, The Transmission of Heat and Gases through, and the Condensation of Moisture on the Surface of, Wall Materials* (1921).

26. NA, HLG 52/881, Astor Committee, 18th meeting, 19 May 1920.

27. NA, DSIR 3/50, BMRC, 39th meeting, 3 June 1920; Building Materials Research Committee, 'Stability of Thin Walls' and 'Heat Transmission through Walls', *Housing* 2 (19 July 1920) pp10–11.

28. S Marriner, 'Cash and concrete. Liquidity problems in the mass production of "homes for heroes"', *Business History* 18 no 2 (1976), pp152–189; Swenarton, *op cit*, p125.

29. NA, DSIR 4/1, Advisory Committee, Memorandum on Building Research, January 1919, and minute of 8 January 1919.

30. NA, DSIR 4/1, memorandum by Curzon, 23 April 1919.

31. NA, DSIR 4/1, letter to Local Government Board, 3 February 1919.

32. NA, CAB 24/92, CP3, memorandum by C Addison, 27 October 1919.

33. NA, DSIR 4/1, minute of Advisory Committee, 5 November 1919.

34. NA, DSIR 4/1, minute of Advisory Committee, 17 December 1919.

35. NA, DSIR 4/1, meeting of sub-committee appointed by the Advisory Council, 26 February 1920.

36. NA, DSIR 4/1, Note on Building Research Board, 3 February 1920, and letter from Salisbury to Heath, 11 February 1920; DSIR 4/4, Building Research Board, 1st meeting, 24 June 1920.

37. NA, DSIR 3/50, 40th meeting of the BMRC, 3 December 1920.

38. NA, DSIR 4/1, Sir M Fitzmaurice to Ogilvie, 9 June 1920.

39. NA, CAB 24/72 GT 6552, memorandum by Sir A Mond, 23 December 1918; CAB 24/93 CP 107, memorandum by Mond, 11 November 1919; CAB 24/107 CP 1455, memorandum by Mond, 14 June 1920; Swenarton, *op cit*, p124.

40. NA, DSIR 4/1, interview with Sir L Earle, 24 June 1920.

41. NA, DSIR 4/1, letter from Sir L Earle, 24 June 1920.

42. NA, DSIR 4/1, letter from HM Office of Works, 16 August 1920.

43. NA, DSIR 4/4, BRB, 1st meeting, 24 June 1920, and 3rd meeting, 25 November 1920; PP 1921 Cmd 1491 xvii, 'Report of the Committee of the Privy Council for Scientific

and Industrial Research for the year 1920–1921', p49. See also Building Research Board, *Special Report no 18, The weathering of natural building stones by RJ Schaffer* (1932).

44. NA, DSIR 4/5, BRB, 4th meeting, 18 February 1921.

45. NA, DSIR4/5, BRB, 3rd meeting, 25 November 1920; 4th meeting, 18 February 1921; and 5th meeting, 15 April 1921.

46. NA, CAB 24/126 CP 3111, memorandum by A Mond, 7 July 1921; NA, CAB 24/126, Amended Draft Statement by the Minister of Health; *Parliamentary Debates, Commons* 144 cols 1483–1485, 14 July 1921 (Sir A Mond); Swenarton, *op cit*, pp133–5; S Marriner, 'Sir Alfred Mond's Octopus: a Nationalised House-Building Business', *Business History* 21 no 1 (1979) pp23–44.

47. NA, DSIR 4/4, BRB, 3rd meeting, 25 November 1920, and Director's Report to 6th meeting, 27 July 1921; Cmd 1491, pp48–49; Lea, *op cit*, p16; BREA, Internal Note IN98/66, pp44–5; Atkinson, *op cit*, p447.

48. NA, DSIR 4/5, BRB, Director's Report to 8th meeting, 14 October 1921; DSIR 4/1, Carmichael to Heath, 26 November 1919.

49. NA, DSIR 4/1, Weller to Salisbury, 4 October 1921, and Forber to Heath, 22 October 1924.

50. NA, DSIR 4/7, BRB, 21st meeting, 28 November 1923; DSIR 4/2, Report by the Building Research Board, nd (December 1923), and Heath to Earle, 3 January 1924.

51. NA, DSIR 4/1, Forber to Heath, 22 October 1924.

52. PP 1924–25 Cmd 2491 xv, 'Report of the Committee of the Privy Council for Scientific and Industrial Research for the Year 1924–1925', pp6 and 55; BREA, Internal Note IN98/66, pp70–75; G Atkinson, *op cit*, p447; Lea, *op cit*, pp17–18; also chapter eleven below.

53. BRS, *Golden Jubilee Congress*, p188.

Chapter eleven – Houses of paper and brown cardboard

1. Neville Chamberlain to his sister Ida, 23 May 1924, in R Self (ed), *The Neville Chamberlain Diary Letters vol 2 The Reform Years 1920–1927* (3 vols, Aldershot, 2002), p225.

2. RB White, *Prefabrication: A history of its development in Great Britain* (1965), pp42–8; FM Lea, *Science and Building: A history of the Building Research Station* (1971), pp10–11; M Swenarton, *Homes fit for Heroes: The Politics and Architecture of Early State Housing in Britain* (1981), pp106–8; also chapter ten above.

3. Lea, *op cit*, p17.

4. In their published accounts of the move to Garston neither White nor Lea mention Chamberlain: White, *op cit*, p67; Lea, *op cit*, p17. A more rounded picture is given in the unpublished 1966 history by RB White, which was one of the sources for George Atkinson's 1971 article on Unwin and the BRS: BREA, Building Research Station, Internal Note IN98/66, *Information Division, A short history of Building Research in the United Kingdom from 1917 to 1946 – in particular that of the Building Research Station, by RB White, with the collaboration of Station Officers*, 1966, pp70–5. See G Atkinson, 'Raymond Unwin: founding father of BRS', *RIBA Journal* 78 (October 1971) pp446–8, which also informed D Hawkes, 'Garden Cities and New Methods of Construction:

Raymond Unwin's influence on English housing practice, 1919–1939', *Transactions of the Martin Centre for Architectural and Urban Studies* 1 (1976) pp275–96. See also F Jackson, *Sir Raymond Unwin: Architect, Planner and Visionary* (1985), pp134–139.

5. NA, DSIR 4/65, memorandum to the Lord President from Sir F Heath, 27 February 1925, p2.

6. GE Cherry, 'Introduction: aspects of twentieth-century planning', in GE Cherry (ed), *Shaping an Urban World* (1980), p13. See also GE Cherry, 'The place of Neville Chamberlain in British town planning', in GE Cherry, *ibid*, pp161–179; and P Hall, 'The Centenary of Modern Planning', in R Freestone (ed), *Urban Planning in a Changing World: The Twentieth Century Experience* (2000), p26. For the Greater London Regional Planning Committee, see M Miller, 'The elusive green background: Raymond Unwin and the Greater London Regional Plan', *Planning Perspectives* 4 no 1 (January 1989) pp15–44. See also S. Pepper, 'Unfit for Heroes: the Slum Problem and Neville Chamberlain's Unhealthy Areas Committee, 1919–21', *Town Planning Review* (forthcoming).

7. C Macintyre, 'Policy Reform and the Politics of Housing in the British Conservative Party 1924–1929', *Australian Journal of Politics and History* 45 no 3 (1999), p418.

8. See S Merrett, *State Housing in Britain* (1979), pp42–3; also S Lowe and D Hughes (eds), *A New Century of Social Housing* (Leicester, 1991), pp6–7.

9. AJP Taylor, *English History 1914–1945* (1965; Harmondsworth, 1970), p303.

10. Building Research Station, *Golden Jubilee Congress: Proceedings 9–15 June 1971* (1972), p211.

11. TL Webb in *ibid*, p225.

12. J Dick, foreword, in DAG Reid, *Construction Principles 1 Function* (1973), p1. See also Department of Scientific and Industrial Research, *Principles of Modern Building 1 Walls, Partitions and Chimneys* by R Fitzmaurice (1938); Department of Scientific and Industrial Research (Building Research Station), *Principles of Modern Building 1* (3rd edition, 1959) and *Principles of Modern Building 2 Floors and Roofs* (1961); and S Groák, *The Idea of Building* (1992), pp xiv and xvi–xvii.

13. See FM Lea and CH Desch, *The Chemistry of Cement and Concrete* (1935); PC Hewlett (ed), *Lea's Chemistry of Cement and Concrete* (4th edition, 1998).

14. Chamberlain to Hilda, 18 May 1924, in Self, *op cit*, p223.

15. *ibid*, p9.

16. *Parliamentary Debates, Commons* 179 1924–25 col 846, 16 December 1924 (N Chamberlain). *Public General Acts* 13 & 14 Geo 5, ch 24, Housing etc Act 1923.

17. Chamberlain to Hilda, 18 May 1924, in Self, *op cit*, p222; J Ramsden, *The Age of Balfour and Baldwin 1902–1940* (1978), pp188–217; J Charnley, *A History of Conservative Politics 1900–1996* (Basingstoke, 1996), pp69–75.

18. Chamberlain to Ida, 1 April 1922, in Self, *op cit*, p105.

19. Self, *op cit*, p19. See also D Dilks, *Neville Chamberlain 1 Pioneering and Reform 1869–1929* (Cambridge, 1984), p407; M Cowling, *The Impact of Labour 1920–1924* (Cambridge, 1971), p405.

20. *The Times* (20 June 1924) p9.

21. RS Hayward, *Housing by Committee: Some aspects of municipal housing in the first decade after the Great War, with special reference to the City of Liverpool* (University of Liverpool PhD, 1983), pp22–29.

22. Parliamentary Papers 1924 Cmd 2104 vii, 'Report on the Present Position in the Building Industry, with regard to the carrying out of a full Housing Programme, having particular reference to the means of providing an adequate supply of labour and materials', p10.

23. Parliamentary Papers 1924–25 Cmd 2450 xiii, 'Sixth Annual Report of the Ministry of Health 1924–25', p53.

24. Chamberlain to Ida, 18 May 1924, in Self, *op cit*, p227.

25. Chamberlain to Ida, 19 July 1924, in *ibid*, pp238–9.

26. J Cornes, *Modern Housing in Town and Country* (1905); M Swenarton, *Homes fit for Heroes: The Policy and Design of the State Housing Programe of 1919* (University of London PhD, 1979), pp302–29.

27. PP 1913 Cd 6708 xv, 'Report of the Departmental Committee appointed by the President of the Board of Agriculture and Fisheries to inquire into and report as to Buildings for Small Holding in England and Wales, together with Abstract of the Evidence, Appendices, and a series of Plans and Specifications', p3.

28. S Pepper and M Swenarton, 'Home Front: Garden Suburbs for Munition Workers, 1915–1918', *Architectural Review* 163 no 976 (June 1976) pp366–375.

29. Ministry of Health. *Standardisation and New Methods of Construction Committee, Report on the First Year's Work of the Committee, April 1919–April 1920* (1920), appendix 6; S Marriner, 'Cash and concrete. Liquidity problems in the mass production of "homes for heroes"', *Business History* 18 no 2 (1976), pp152–89; Swenarton, *op cit* (1979), pp321–9.

30. Self, *op cit*, p24. For an account of the 'mass production' techniques used in the manufacture of the Weir house, see PP 1924–25 Cmd 2392 xiii, 'Report by a Court of Inquiry concerning Steel Houses', pp17–18.

31. Chamberlain to Hilda, 24 March 1923, in Self, *op cit*, p155. See also NA, DSIR 4/69, Ministry of Health, Committee on New Methods of House Construction, 'Inquiry by the Committee on New Methods of House Construction held at Messrs G & J Weir's Works, Cathcart, Glasgow, on Saturday 4th October 1924', p2.; and K Middlemas (ed), *Thomas Jones: Whitehall Diary 1 1916–1925* (1969), p229.

32. NA, DSIR 4/69, 'Inquiry.... on Saturday 4th October 1924', p3.

33. *Parliamentary Debates, Commons* 179 cols 858–860, 16 December 1924 (N Chamberlain).

34. The attempt by local authorities to build Weir houses without observing union rates triggered a conflict with the NFBTO, to which Chamberlain responded with a Court of Inquiry (March–April 1925) and a subsequent announcement that the government would build 2000 steel (Weir, Atholl and Cowieson) houses in Scotland without involving local authorities. See R White, *op cit* (1965), pp73–77; Self, *op cit*, p25 and pp272–3 et seq; Cmd 2392, *op cit*.

35. Dilkes, *op cit*, p263; D Jarvis, 'Mrs Maggs and Betty: the Conservative Appeal to Women Voters in the 1920s', *Twentieth Century British History* 5 no 2 (1994) pp129–52.

36. Chamberlain to Ida, 8 February 1925, in Self, *op cit*, pp269–70.

37. Hayward, *op cit*, p31. See also IS Wood, *John Wheatley* (Manchester, 1990), pp136–7.

38. *Bills Public* 1924, ii, Bill 167, Housing (Financial Provisions), 5 June 1924.

39. *Public General Acts* 14 & 15 Geo 5 ch 35, Housing (Financial Provisions) Act, 1924, cl 10 and cl 2. See also *Parliamentary Debates, Commons* 176, Housing (Financial Provisions) Bill in Committee, col 1044, 21 July 1924 (J Wheatley) and col 1045, 21 July 1924 (N Chamberlain); *Bills Public* 1924 ii, Bill 248, Lords Amendments to the Housing (Financial Provisions) Bill, 5 August 1924, p353.

40. PP 1924–25 Cmd 2450 xiii, 'Sixth Annual Report of the Ministry of Health', p49.

41. NA, DSIR 4/69, Ministry of Health Committee on New Methods of House Construction, minutes of 4th meeting, 10 October 1924; Interim Report, 4th November 1924. PP 1924–25 Cmd 2310 xiii, Second Interim Report of the Committee on New Methods of House Construction, 7 January 1925. At the first meeting of the committee, Col Levita of the London County Council warned that, due to 'the prejudice which existed in the mind of the ordinary tenant against new methods', the houses would not command the same rent as ordinary houses, so there would be no saving for the local authority: NA, DSIR 4/69, Ministry of Health Committee on New Methods of House Construction, minutes of first meeting, 24 September 1924.

42. NA, DSIR 4/1, Building Research Board, ER Forber to Sir F Heath, 22 October 1924; also chapter ten above.

43. Middlemas (ed), *op cit*, p303, 8 November 1924.

44. K Feiling, *The Life of Neville Chamberlain* (1970), pp459–62; Macintyre, *op cit*, p411.

45. I Dale (ed), *Conservative Party General Election Manifestos* (2000), p34.

46. Middlemas (ed), *op cit*, p303, 8 November 1924.

47. In July 1926, following the failure of the General Strike, Chamberlain announced a reduction from £6 to £4 for the 1923 Act subsidy and from £9 to £7 10s for the 1924 Act subsidy. In 1928 he announced that the 1923 Act subsidy would be abolished and the 1924 Act subsidy further reduced, with effect from September 1929, although this latter was countermanded by the incoming Labour government in 1929. The 1924 Act subsidy was finally abolished by the 1933 Housing Act. See M Bowley, *Housing and the State 1919–1944* (1945), pp45–6; Merrett, *op cit*, pp48–9 and p55.

48. See *Parliamentary Debates, Commons* 179 col 927, 16 December 1924.

49. Chamberlain to Hilda, 15 November 1924, in Self (ed), *op cit*, p260.

50. Middlemas (ed), *op cit*, p307, 28 November 1924.

51. Churchill to Baldwin, 28 November 1924, quoted in Dilks, *op cit*, p422.

52. *ibid*, p415.

53. *Parliamentary Debates, Commons* 180 col 480, 12 February 1925 (Sir K Wood). Needless to say, the Weir houses cost more to build than Weir had predicted and, with general building costs falling, by the middle of 1926 Chamberlain was losing interest in them. See Chamberlain to Ida, 20 June 1926, in Self (ed), *op cit*, p354.

54. *Parliamentary Debates, Commons* 179, col 853, 16 December 1924 (N Chamberlain).

55. *Parliamentary Debates, Commons* 179, cols 858–860, 16 December 1924 (N Chamberlain).

56. Cmd 2310, *op cit*, and PP 1924–5 Cmd 2334 xiii, 'Third Interim Report of the Committee on New Methods of House Construction', 29 January 1925.

57. Cmd 2310, *op cit*, p1.

58. NA, DSIR 4/65, DSIR Advisory Council 25 February 1925, Memorandum on a Request from the Ministry of Health for Research of Direct Application to the Housing Problem, p1.

59. *Parliamentary Debates, Commons* 180 col 487, 12 February 1925 (T Thompson) and col 561, 13 February 1925 (Lt Commander CD Burney).

60. See chapter ten above.

61. NA, DSIR 4/65, memorandum to the Lord President from H Frank Heath, 27 February 1925, pp2–3. For New Delhi see DSIR 4/7, 16th meeting of the Building Research Board, 9 February 1923.

62. NA, DSIR 4/63, Treasury Authority for Appointment of a New Director, 1924; BREA, Internal Note IN98/66, p52.

63. NA, DSIR 4/9, Building Research Board, minutes of 27th meeting, 9 January 1925.

64. NA, DSIR 4/65, Memorandum on a Request, p1. See also *Parliamentary Debates, Commons* 180 col 520, 12 February 1925 (N Chamberlain).

65. NA, DSIR 4/65, memorandum to the Lord President from H Frank Heath, 27 February 1925, p1.

66. NA, DSIR 4/65, Memorandum on a Request, appendix I, letter from ER Forber, 18 February 1925.

67. NA, DSIR 4/65, DSIR Advisory Council, Extract of minutes of meeting of 25 February 1925.

68. NA, DSIR 4/65, Memorandum on a Request, p2.

69. NA, DSIR 4/65, memorandum to Lord President from H. Frank Heath, 27 February 1925, p4.

70. NA, DSIR 4/1, memorandum by Lord Curzon, 23 April 1919; see above, chapters nine and ten.

71. NA, DSIR 4/65, handwritten note by Lord Curzon (erroneously marked '2.4.25'), 2 March 1925

72. NA, DSIR 4/65, minute of BRB meeting of 24 March 1925.

73. NA, DSIR 4/65, letter from HM Office of Works to the Treasury, 9 May 1925.

74. NA, DSIR 4/65, letter from DSIR to HM Office of Works, 17 November 1925.

75. NA, DSIR 4/98, BRB chairman Sir Gerard Heath to RE Stradling, 11 December 1925.

76. NA, DSIR 4/10, BRB, minutes of meeting of 30 April 1926, p1. See also DSIR, *Report of the Building Research Board for the Period Ended 31st December 1926* (1927).

77. NA, DSIR 4/65, Memorandum on a Request, appendix II, Programme of Special Investigation Work to Assist in the Housing Shortage, pp4–5.

78. NA, DSIR 4/65, letter from the Treasury to the DSIR, 7 April 1925.

79. NA, DSIR 4/70, DSIR, Proposals for Research in Buildings, 6 March 1925.

80. NA, DSIR 4/130, Paper for discussion between representatives of the Ministry of Health, Office of Works and Director of Building Research on suggested expansion of work to assist on housing work, pp1–19.

81. NA, DSIR 4/130, Unwin to Stradling, 15 May 1925. For Unwin's visit to New York, see M Miller, 'Transatlantic Dialogue: Raymond Unwin and the American Planning Scene', *Planning History* 22 no 2 (2000) p20.

82. NA, DSIR 4/130, Unwin to Stradling, 15 May 1925. The search for an alternative to plaster and, more especially, plasterers remained the priority for the Ministry throughout 1926, with the BRS investigating both walling boards and plaster guns as possible substitutes. But before a satisfactory solution could be found the shortage of plasterers had become less acute and by 1928, at Unwin's instruction, the work was set aside. See NA, DSIR 4/98, Heath to Stradling, 27 January 1926, and Stradling to Heath, 28 January 1926 and 1 October 1926; also NA, DSIR 4/130, Building Research Station, Notes on Progress of Work, nd (August 1929).

83. NA, DSIR 4/130, Unwin to Stradling, 15 May 1925.

84. NA, DSIR 4/130, Memorandum of Discussion at the Ministry of Health on the Housing Research Programme, 22 May 1925.

85. NA, DSIR 4/130, R Unwin, Notes of informal discussion, 23 May 1925.

86. Lea, *op cit*, pp17–18.

87. DSIR, *Report of the Building Research Board* (1927), p8.

88. DSIR, *Report of the Building Research Board for the year ended 31st December 1929* (1930), p8; Lea, *op cit*, p30.

89. Lea, *op cit*, p24.

90. Self, *op cit*, pp24–26. In total, about 1700 Weir houses were built between 1925 and 1927, mostly (1552) in Scotland. See White, *op cit* (1965), p77; H Harrison, S Mullin, B Reeves and A Stevens, *Non-traditional houses: Identifying non-traditional houses in the UK 1918–1975* (Watford, 2004), p912; Scottish Executive, A *Guide to Non-Traditional Housing in Scotland* (Norwich, 2001), p14. White estimated the overall number of non-traditional houses built up to 1928 at 50,000, equivalent to just over 12 per cent of the 411,900 local authority houses built in England and Wales in the period. See White, *op cit* (1965), p88; Bowley, *op cit*, p272.

91. See B Finnimore, *Houses from the Factory: System Building and the Welfare State 1942–1974* (1989).

92. Lea, *op cit*, pp195–8; BRS, *Golden Jubilee Congress: Proceedings, op cit*, pp225–6.

Index